www.ingramcontent.com/pod-product-compliance
Lightning Source LLC
Chambersburg PA
CBHW051251200926
51612CB00001B/3
9780739125991

Reproducing Chinese Culture in Diaspora

Reproducing Chinese Culture in Diaspora

Sustainable Agriculture and Petrified Culture in Northern Thailand

Shu-min Huang

LEXINGTON BOOKS

A division of

ROWMAN & LITTLEFIELD PUBLISHERS, INC.
Lanham • Boulder • New York • Toronto • Plymouth, UK

Published by Lexington Books
A division of Rowman & Littlefield Publishers, Inc.
A wholly owned subsidiary of The Rowman & Littlefield Publishing Group, Inc.
4501 Forbes Boulevard, Suite 200, Lanham, Maryland 20706
http://www.lexingtonbooks.com

Estover Road
Plymouth PL6 7PY
United Kingdom

British Library Cataloguing in Publication Information Available

Library of Congress Cataloging-in-Publication Data

Huang, Shu-min.
Reproducing Chinese culture in diaspora : sustainable agriculture and petrified culture in Northern Thailand / Shu-min Huang.
p. cm.
Includes bibliographical references and index.
ISBN 978-0-7391-2599-1 (alk. paper)—ISBN 978-0-7391-4344-5 (electronic)
1. Sustainable agriculture—Thailand. 2. Chinese—Thailand. 3. Chinese—Foreign countries. I. Title.
S494.5.S86H837 2010
338.1089'9510593—dc22 2009032408

Printed in the United States of America

∞™ The paper used in this publication meets the minimum requirements of American National Standard for Information Sciences—Permanence of Paper for Printed Library Materials, ANSI/NISO Z39.48-1992.

Contents

List of Illustrations

Acknowledgments

The idea of conducting a research project in Maehong began in summer 2000 when I met Professor Hui Yang of the Department of Anthropology, Yunnan University, at an international conference in Xiamen, Fujian. Professor Yang and her late husband, Professor Zhusheng Wang, also an anthropologist at Yunnan University, along with some of their graduate students, had conducted ethnographic fieldwork on Yunnan Chinese in Maehong in early 1990s. When Professor Wang passed away, Professor Yang felt devastated and stopped the fieldwork for a number of years. When I met Professor Yang in Xiaman, she invited me to join her in the project and asked me to look for possible funding to revive fieldwork in this village.

After studying extant publications about the contorted history of the Yunnan Chinese diasporas in general and the complex interethnic relations in the Golden Triangle, I became fascinated, from both theoretical and practical points of view, and formulated a funding proposal to carry out fieldwork in this particular community, Maehong. A research grant from the Wenner-Gren Foundation for Anthropological Research allowed me to launch the project in winter of 2002.

Over the years doing fieldwork in Maehong, I have accumulated tremendous personal debts to people and organizations that have made this project possible. In Maehong village, my principle host, Mr. Huang and his family, have always been most gracious in receiving me and enduring my endless questions. Mr. Huang, in many of our late evening drinking sessions, provided the overall picture about Maehong and its specific incidents that I used as the framework of the current book. A second host, Mr. Liu, has also generously taken me in as part of his family. The daily trip in his beat-up Toyota pickup truck to his hill farms was always the most enjoyable and enlightening occasion, and taught me a great deal about the plants, animals, and the people in this area. From Mr. Liu, I learned a great deal about the suffering of the Chinese youth who escaped China during the Cultural Revolution era. Other villagers, such as Mr. Peng and his family, M. Yang (the former school principal) and his family, and Mr. Duan (the Chinese school principal in 2007), all made their best effort to accommodate me when I visited Maehong.

A number of colleagues have been instrumental in expanding my knowledge and understanding about issues and problems in the field. Professor Hui Yang provides the main entry for me and other team members so that we would be accepted by Maehong villagers. Without her introduction, we would have all faced daunting challenges in the field. Professor Mattiga Panomtarinichagul, a soil scientist from Chiang Mai University, was a gracious host whenever I visited Chiang Mai. Watching her arduously climbing up and down the hills collecting soil samples made me realize the close affinity between anthropology and biological sciences. Mr. Duan, a PhD candidate at the Department of Anthropology, Chinese University of Hong Kong, was a good companion in the field. Since we shared a room in Maehong, we engaged in many late night discussions about the current state of anthropology and its development trends. Mr. Duan also often let me copy his field notes when I was unable to do so due to my official position in formal meetings. Professor Pong-In Rakariyatham, a geographer and vice president of Chiang Mai University, although unable to attend as many actual field trips as he wished due to his busy official schedule, managed to meet all researchers either in the field or in Chiang Mai whenever possible to learn about the latest project development.

Two research grants from Wenner-Gren Foundation for Anthropological Research under its International Collaborative Research Program (2002–03 and 2004–05) provided the main financial backbone for this project. I am most grateful to Professor Richard Fox, then president of the Foundation, and Ms. Pamela Smith, project official for International Collaborative Research Program, for their generosity. Additional funding for 2003–04 came from Iowa State University's Bridge Grant. I appreciate all these institutions' support that was indispensable for the completion of the project in a timely fashion.

To safeguard the anonymity of the village and its residents, I used pseudonyms for all, except for the public figures whose names appeared in print elsewhere.

Chapter 1

Introduction: The Golden Triangle and the Yunnan Chinese in Northern Thailand

This is a story about a group of armed Chinese refugees who have managed to establish a rural livelihood in Thailand's northern hills over the past half century. My study concerns not only these refugees and their hardships as they struggled to carve out this new life, but also covers related issues—the practicality and sustainability of the hill farming they developed, and the implications of this livelihood for environmental conservation, drug eradication, cultural reproduction, and the persistence of Chinese ethnicity in a transnational matrix. Thus, my ethnographic narrative is simultaneously descriptive, analytical, theoretical, and policy-oriented.

This particular Chinese diasporic community mostly consisted of Nationalist military personnel and their dependents from southwestern China's Yunnan Province. When faced with the seemingly invincible Communist forces that were sweeping across the Mainland, these Yunnan Chinese retreated in 1949 or shortly thereafter to the Golden Triangle area that sits astride the borders of Burma, Laos, and Thailand. Their exodus from China was motivated primarily by a brand of radical nationalism instigated by the Chinese Nationalist Party (popularly known as Kuomintang, or KMT), which perceived communism to be an evil political doctrine imported from the West for the explicit purpose of destroying traditional Chinese culture, specifically Confucian orthodoxy.

As defenders of this precious cultural heritage, the Nationalists were willing to endure all manner of harsh conditions, including engaging in military campaigns, to resist the Communist forces. Upon being defeated, they abandoned their ancestral homes to pursue a precarious and difficult living in the tropical jungles of the far southwest; they suffered greatly for their cause, barely surviving. While their vehement anti-communist ideology sustained

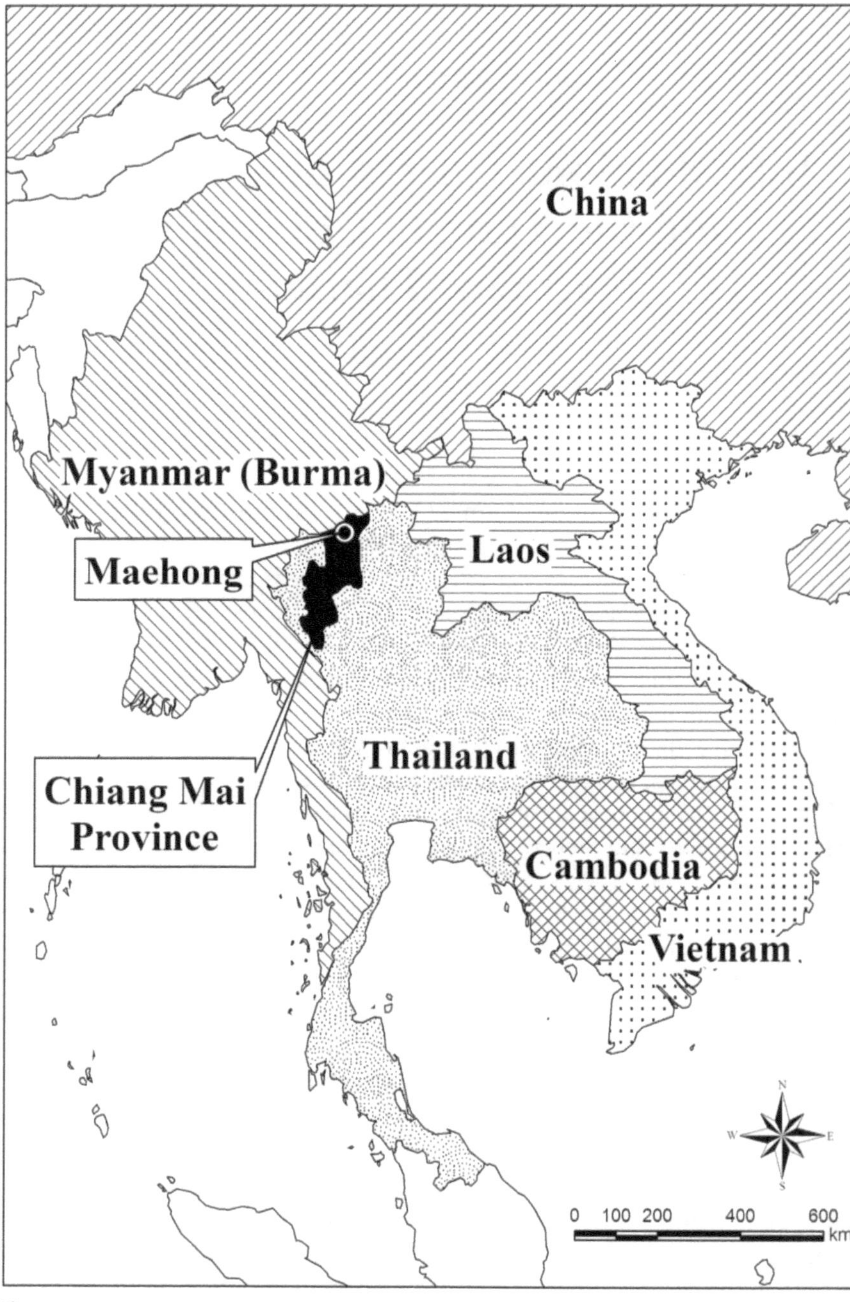
China
Myanmar (Burma)
Laos
Maehong
Thailand
Chiang Mai
Province
Cambodia
Vietnam
N
W
E
S
0 100 200 400 600
km

Figure 1.1.

them in their struggles, this same moral commitment also led them to engage in the morally objectionable activities of opium cropping, the production of heroin, and international drug trafficking. In the name of anti-communism, they justified their conduct by pointing out the necessity of drug involvement for their survival.

As a disciplined armed force, the Yunnan Chinese quickly became dominant "drug lords" in the Golden Triangle and remained so throughout the 1950s. Under pressure from both the Burmese and Chinese military, the Nationalist government in Taiwan resettled the most of its forces from Burma to Taiwan in two waves, in 1953 and 1961. In the early 1960s remnants of the KMT forces retreated further south and entered Thailand's northern hills where they established several permanent refugee communities. Having helped the Thai army eradicate several local communist insurgent groups in the north, they were granted in return formal permission to live in their newly adopted country, eventually receiving Thai citizenship in 1981.

With their new resident status, the Yunnan Chinese completely changed their *modus operandi* and established a hill farm system that produces tropical fruit in perennial orchards. By experimenting with various fruit tree varieties and selling their produce on domestic and international markets, the Yunnan Chinese have brought a degree of prosperity to this region—as can be seen in their manicured orchards, colorfully furbished houses, and well-stocked grocery stores. In this way, they have been able to gradually shed the notoriety they gained because of past involvement in drug production and trafficking.

These hill farms may appear sound and viable, but their communities' capacity to renew themselves and preserve their sociocultural identity remains in question. My inquiry thus focuses on the extent to which the Yunnan Chinese in fact *have* succeeded in building a sustainable agricultural regime that is economically viable, environmentally friendly (relatively speaking), and supportive of sociocultural self-renewal. Furthermore, I interrogate the specific sociocultural factors that have contributed to their successes and failures.

My reservations about the sustainability of this new rural livelihood do not derive from the technical or environmental aspects of the hill farming system. In fact, this agricultural regime the Yunnan Chinese established has provided clearly demonstrated alternative to opium production that contains identifiable replacement cultigens, operational techniques, and environmentally sound practices—within the local community scale and of short-term duration. It could be a model for other hill farmers in the Golden Triangle who wish to escape the dual scourges of poverty and drug involvement. In this sense the Yunnan Chinese have demonstrated unequivocally that hill farming need not be environmentally detrimental due to deforestation or soil erosion, nor must it be unprofitable to the farmers. This model should receive serious consideration by the various national or international agencies engaged in anti-drug and anti-poverty campaigns in the region.

And yet the successes of the Yunnan Chinese in building a viable rural livelihood are not unqualified. As will be explained in subsequent chapters, the paradox lies embedded in their entrenched cultural heritage. On the one hand, the Chinese cultural repertoire provides the necessary social and cultural capital (such as knowledge, skills, management practices, market information, and so on) with which to handle practical problems of this farming system and make it both environmentally and economically viable. And yet, it is exactly this same set of cultural ideals—which they have struggled arduously to preserve and reproduce—that has rendered the community inflexible, self-negating, and hence unsustainable, a process I call "cultural petrification."

By petrification I mean when certain cultural ideals or practices become sacrosanct and cannot be questioned, challenged, or altered. Established cultural forms and procedures are revered more than concerns about substance or meanings—a condition that stifles creative thinking and innovation.[1] Among the Chinese communities in northern Thailand, this sort of inflexibility has driven away the younger generations in droves, and those who stay behind are in constant rebellion against the self-proclaimed cultural authorities. This incessant conflict and division casts a dark cloud over the future of these diasporic communities.

Further complicating these problems are factors that extend far beyond the local Yunnan Chinese. They include the intermittent civil wars in neighboring Myanmar (formerly Burma) where the hill farmers go for cheap labor; the volatile domestic politics of the Thai nation-state, which has made some of the Chinese hill farmers' legal status in Thailand precarious; the shifting social, political, and economic conditions in Taiwan and China from where the Yunnan Chinese draw their nationalistic inspiration and cultural identity; and, finally, changing global politics which, during the Cold War, made the Golden Triangle a major opium production center, but with the anti-terror campaigns after 9/11 has shifted the focus of opium production to war-torn Afghanistan.

Thus, if the history of the Yunnan Chinese community in the Golden Triangle has meaning on multiple levels, so do the theoretical implications we may extrapolate from it. My inquiry delves into three problem domains: (1) the short- and long-term environmental impact this agricultural regime will have on the hill country ecosystem; (2) the significance of Chinese cultural identity and ethnicity in this agro-ecological adaptation process, and (3) the multiple and changing identities of the diasporas and its significance in the emerging global community. To carry out this project, I assembled a multidisciplinary (i.e., anthropology, geography, filming, and soil science) and multinational (i.e., China, Thailand, and the U.S.) research team and successfully obtained necessary funding for its operation.[2] From late 2002 through the summer of 2005, these researchers—who employed both qualitative and quantitative approaches in data collection—conducted fieldwork in the village of Maehong.

Let me proceed with a short description of general conditions in the Golden Triangle of northern Thailand, to be followed by discussion of the composition and the general characteristics of the Chinese Nationalist soldiers who left Yunnan to enter this region. After this brief introduction, I will discuss the development trajectory of the village of Maehong, the organization of this book, and some of the initial findings that provide tentative answers to my research questions. It is hoped that in doing so the readers will gain a general understanding of how we conducted our research and the significance, both theoretical and practical, of our project.

THE REGION OF THE GOLDEN TRIANGLE

The Golden Triangle straddles the three nation-states of Thailand, Myanmar (formerly Burma), and Laos; it covers approximately 200,000 square kilometers and has dramatic topographic features, including major rivers, rugged mountains, lowland basins, and river valleys (Anderson1993; Forbes and Henley 1997; Geddes 1983; Kunstadter 1983; Le Bar et al., 1964; Lewis and Lewis 1984; Li 2001a, 2001b; Santasombat 2003; Young 1962). This fragmented landscape is both ethnically and biologically diverse. Its stratified human adaptation pattern (or vertical ethnic hierarchy) consists of: the Thais and Shans in the lowest basins, who construct paddy fields for subsistence farming; the Karens, inhabiting the valleys above them, who also construct terraces for long-term wet rice cultivation.

Farther up are the Lisu, Wa, Miao (also called Hmong or Meo in Vietnam and Laos), and Yao (or Mien in Laos), who occupy the middle girdle of the hills, where they plant seasonal crops such as dry rice, yams, corn, and sweet potatoes. With simple agricultural technologies, they maintain medium-term fallowing cycles with relatively stable yields. Opium growth requires high elevations, at least 1,000 meters above sea level, and historically this is a major reason some of these groups preferred to live in higher elevations. Higher up still are swidden agriculturalists Akha and Lahu, who also rely on hunting and foraging to supplement their agriculture. Using relatively simple agricultural technologies they plant hill rice and other subsistence crops (Anderson 1993; Geddes 1976; Kunstadter 1983; Lewis and Lewis 1984; Sturgeon 2005; Young 1962).

In northern Thailand, the Golden Triangle covers parts of the three provinces of Chiang Mai, Chiang Rai, and Mae Hong Son, an area of approximately 128,480 square kilometers. More than 80 percent of this area is mountainous. Besides the rice terracing of the Thais and Karens, all other slash-and-burn agriculture in the hills is considered extractive in nature, and these agricultural practices are deemed detrimental to the environment by officials in the Royal Forestry Department (Anderson 1993; Geddes 1976; Li 2001a, 2001b; Kunstadter 1983; Rerkasem et al. 1994; Santasombat 2003). With continuous population growth and lacking an alternative means of making a living in the lowland areas, land-hungry farmers have pushed up the slopes and cut the traditionally long fallow period of 5–10 years down to 1–2 years; some have completely eliminated it in favor of intensive farming (Geddes 1976; Kunstadter 1983; Lewis and Lewis 1984; Ongprasert and Turkelboom 1999; Rerkasem et al. 1994; Santasombat 2003; Sturgeon 2005). Because their low levels of productivity contribute to general poverty, many hill tribes regarded poppy farming as an attractive way to supplement their subsistence crops (Feingold 2000; Rerkasem et al. 1994; Sturgeon 2005).

To counter poverty, deforestation, environmental degradation, and opium production, many agencies and organizations, both domestic and international, have established experimental stations and extension services in this region, hoping to find substitute cultigens that are economically viable, environmentally friendly, and locally acceptable (Rerkasem et al. 1994; Santasombat 2003). Newly introduced crops include wheat, buckwheat, barley, soybeans, cabbage, chilis, and potatoes, which have been promoted as cash crops, and hedge-grass (Vertivers, also called the King's grass because it was endorsed by the royal family-sponsored projects) to prevent soil erosion. Despite these investments, most experiments have seen limited success. The dual problems of poverty and environmental degradation persist (Anderson 1993; Geddes 1983; Li 2001a, 2001b; Panomtarinichigul, et al., 2002; Rerkasem et al. 1994).

YUNNAN CHINESE IN NORTHERN THAILAND

Yunnan Chinese have long made their presence felt in northern Thailand, especially the Muslim caravan traders who pursued transnational trade in Laos, Thailand, Burma, and Tibet (Chang 2001, 2002; Forbes and Henley 1997; Hsieh 1997; Maxwell-Hill 1983; 1998; Mote 1967; Walker 1999;

Young 1962). With porous national borders thanks to lax governmental regulation and sparse border patrols, people in this borderland could easily crisscross the oftentimes unmarked political demarcations by simply wading across a brook, hiking over a mountain ridge, or moving across the street from one household to another (Sturgeon 2005).

The Yunnan Chinese are called Ho, Haw, Chin Ho or Chin Haw by the Thais and other hill tribes. But since these terms are closely associated with the Muslim caravan traders, they are not really appropriate autonyms for Nationalist Chinese who settled in northern Thailand after 1960 (Chang 2001, 2002; Forbes and Henley 1997; Hsieh 1997; Maxwell-Hill 1983; 1998; Mote 1967; Young 1962).[3] For reasons of clarity, then, I use terms such as Yunnan Chinese, Nationalist Chinese, or Yunnan Chinese diasporas in this book and avoid the locally used terms completely.

There are two popular non-fiction works in Chinese about the Nationalist Chinese who exited China and settled in the Golden Triangle after 1949. The first, written and published in Taiwan in the 1960s by the famous dissident writer Bo Yang,(柏楊), is titled *Yiyu* (異域, *Alien Lands*, Bo 1961). The second, by China's well known investigative reporter, Deng Xian (鄧賢) in 1990s, is titled *Lilang jingshanjiao* (流浪金三角), *Wandering in the Golden Triangle*, Deng 2000). These books have been popular with Chinese communities around the world not only because of the excellent writing, but also because they vividly depict a group of Chinese who survived a devastating civil war and the harsh conditions of jungle living in the Golden Triangle to eventually established a new way of life in this exotic new land.[4]

Chronologically, these books cover an almost seamless sequence of events, with Bo Yang's book focusing on the role played by the Nationalist Eighth Field Army in Yunnan during the last years of the civil war (circa 1945–49), when the Nationalists were defeated by Communist forces and driven out of China. Bo Yang depicts how the remnant Nationalist forces, under the command of General Li Mi (李彌), entered the Golden Triangle and regrouped to become a formidable military presence in the northeast Burmese Shan states, developments that took place from approximately 1949 through the early 1960s. He recounts the many heroic actions of Nationalist soldiers and officers who faced adverse conditions and describes their struggles to make ends meet in this untamed land. He hints at, but does not elaborate on, the drug production and drug trafficking conducted by high-level officials among the armed refugees.

Deng Xian's book picks up the narrative from *Alien Lands* and depicts how the Nationalist forces established themselves in the Golden Triangle through active engagement in opium production and trafficking, as well as describing their numerous battles with local drug lords and the Burmese military.

Deng explores in detail the subtle role these military refugees played in the regional political strategies of the Nationalists in Taiwan, the newly funded Chinese Communist regime, the anti-communist Thai Government, and the CIA during the Cold War period. After being abandoned by both Taiwan and the CIA in the early 1960s, this group was left with few options and retreated to Thailand's northern hills where they established several permanent refugee communities.

Once in northern Thailand, these Nationalist units simply expanded their drug activities with a new collaborator—corrupt Thai officials (Cowell 2005; Forbes and Henley 1997; McCoy 1972). As Ronald Renard observed, "When the KMT regrouped in Thailand after 1961 to form the Third and Fifth Armies under Gen. Li Wen Huan (a.k.a. Li Wen-huan) and Gen. Tuan Shi Wen (a.k.a. Duan Xi-wen), respectively, KMT involvement in the opium trade continued. The KMT also became involved in exporting other Burmese commodities such as the excellent jadeite of Kachin State. The KMT thus used their forced resettlement to expand into, and later dominate, the newly emerging Burmese black market that developed after Ne Win's 1962 coup. This enables the KMT to support over 3,000 troops" (Renard 1996, 54–55).

One point that needs to be clarified here is the nature of the Chinese Nationalists who entered the Golden Triangle after 1949. Most of the military refugees were regulars in the Nationalist Army, but a large minority of them was not. In the early 1950s when General Li Mi (李彌), under the auspices of the Nationalist government in Taiwan, set up a military training base called the Anti-Communism University in Tachilek, Burma, it actively recruited from local militias in western Yunnan. Western Yunnan Province is home to many ethnic minorities who were at that time under either tribal chiefdoms or a nominal headman system (*tusi* 土司) appointed by Chinese governing entities, such as the imperial Qing court (circa 1644–1911) or the Nationalist government (1911–49). In this borderland between China's civil administration and foreign polities, local chiefdoms and headmen often maintained private armies or militias for self-protection and law enforcement. Discontented with the newly triumphant Communist regime and its anti-traditional policies, militia groups from western Yunnan left *en masse*—with their original leadership structure intact, to join the anti-communist coalition in the Golden Triangle.

Later, when Burma and China forced the evacuation and resettlement of the main body of the Nationalist military forces to Taiwan (1953–54, with an estimated 6,750 evacuees, and 1961, with another 4,406; see Chang 2001, 2002), those who refused to be evacuated retreated into northern Thailand. Most of these were not army regulars, but their recruits, the regional Yunnan militias; only a small minority was Nationalist regulars.

The refugee groups who moved to Thailand coalesced into two military units, the Third Field Army, centered around the village of Maehong in Chiang Mai Province, and the Fifth Field Army, headquartered in Mae Salong in Chiang Rai Province. Even though they were called field armies, their actual numbers were rather small: the Third Field Army had an estimated 1,700 troops, and the Fifth Field Army slightly over 1,500 (Chang 2001, 2002).[5]

The nature of the local militia organizations had significant impacts on the future growth of these Chinese diasporic communities. Take the Third Field Army as an example. This militia came predominantly from Zhengkang County (鎮康縣) immediately northeast of the Burmese border. General Li Wen-huan (李文煥), head of the militia who was nicknamed "the Mr. Big of Opium" (Cowell 2005: 9), came from the richest family in that county and had led the militia group as if it were his own private force from the early 1940s (Chang 2001, 2002; Cowell 2005; Forbes & Henley 1997; McCoy 1972). Most officials and foot soldiers in the Third Field Army were either General Li's family servants or tenants whose allegiance was primarily to General Li, and only secondarily to the Nationalist government in Taiwan. They maintained an absolute loyalty to General Li and regarded him as the paternalistic authority whose word was the law of the land.

Besides these regionally bounded militias, two other groups joined the Nationalists' anti-communist coalition in the Golden Triangle in 1950s and '60s. The first is the Yunnan caravan traders (but not Muslims) who served as middlemen or procurement agents between the military and the local people. They took orders from the military commands and acquired daily necessities from local traders or whatever sources they could cultivate. The symbiotic relationship between these caravan traders and the military encouraged the traders to accept the strict authority exercised by top officers such as General Li Wen-huan. Even though this group was small, their pecuniary acumen, good management skills, and extensive local knowledge made them indispensable to the top brass. They would eventually play critical roles when the military units established permanent communities in northern Thailand.

A second group who joined the anti-communist coalition was the Burmese Chinese population from the Shan states. A large number of Yunnan Chinese migrated to northeast Burma in late 19th and early 20th century when the national boundaries between China and Burma were vague, especially around the Kokang area (Chang 2001, 2002; Forbes & Henley 1997). These Burmese Chinese had carefully maintained their Chinese cultural traditions and ethnic identity through the establishment of Chinese schools; they also made frequent visits to their ancestral homes in Yunnan. They joined the anti-communist coalition partly out of ideological conviction and partly because they saw joining the military as a possible avenue for social mobility. They have had a lower level of allegiance toward traditional authorities such as General Li Wen-huan and have tended to be more critical of abuses and corruption than the other two groups.

In 1953–54 and again in 1961, when the Nationalist government in Taiwan evacuated its regular troops from the Golden Triangle, it also promised to resettle its allies, namely the Yunnan militias and the Burmese Chinese recruits, to Taiwan. Many in these ally groups refused to be evacuated because of several factors, according to my key informant, Mr. Liu. First, they believed that since they were not regulars of the Nationalist military, they would face discrimination in Taiwan. Second, they wished to be close to their ancestral homes in western Yunnan and northeast Burma. Taiwan was too far away for them to contemplate going home easily. Lastly, many of them were opium addicts and the supply of the drug was abundant in the Golden Triangle. They knew they would face far more difficulty in maintaining their habit if they were to move to Taiwan.

When the two remnant Nationalist militia groups retreated to northern Thailand in 1964, they were actively supported by the Thai military. Having a group of armed settlers on its northern frontier directly facing Burma—the historical enemy of the Thai court (Thongchai 1994)—created a buffer between the two adversaries. Villagers in Maehong today still reminisce nostalgically about the close cooperation they enjoyed with the Thai military. The Thai army sent bulldozers and other earthmoving equipment to clear the original forest so they could build houses and schools, and also built roads for easy transportation.

Since Burma was in these years engulfed in its own incessant civil wars, the armed Chinese settlers had to provide only limited protection of Thailand's northwest frontier; nonetheless, they provided a different service that would enhance Thai national security in later years—the elimination of insurgent communist forces in the north, especially among the Miao. The Thai military recognized the military capability of the Yunnan Chinese settlers early on, despite their prevalent opium addiction problems and rag-tag appearance. Yunnan Chinese military personnel were called upon to engage in two anti-communist campaigns in 1970–74 and 1981, and after the successful 1981 campaign, the Thai government granted citizenship to them and their immediate families.

Having gained Thai citizenship, the Yunnan Chinese now had the legal right to possess land in perpetuity. They purchased large expanses of mountain slope land from the neighboring Lisu, Karen, and Thais, who considered the hills to be of low agricultural value. Conscious efforts were made to increase land productivity by investing in new technology as well as in experimenting with new cultivars. Equally important were the efforts to preserve their newly acquired property through refined land management and farming techniques. Within two short decades, these Chinese established a new hill farming regime that seems to have provided them a viable living in this otherwise inhospitable area.

Another major development that has had immediate impact on the Yunnan Chinese and their successful adaptation in northern Thailand occurred in 1980s: the resumption of contact with the Nationalists in Taiwan. When the Yunnan militias refused to be settled in Taiwan in 1964, the Nationalist government severed all formal relations with them, save for admitting the children of refugee families to schools in Taiwan. In the 1980s Taiwan was experiencing rapid industrial growth and enjoyed a fattened national coffer. News of the abandoned Nationalist soldiers, albeit irregulars, in northern Thailand reached Taiwan in these years, and such stories aroused public sympathy and calls for humanitarian assistance to help them.

The Mainland China Relief Assistance Office set up a separate Northern Thailand Operation Unit to channel assistance to the region via three separate programs. The first program acknowledged these militia units as affiliates of the Nationalist Army; in so doing, all those enlisted became entitled to the retirement pensions given to regular military personnel in Taiwan. Besides the direct monetary allowance, the Nationalist government also provided medical services and health care to the aging militiamen. For instance, in 1988, the Nationalist government built a retirement home in Maehong at a cost of 1.85 million baht (or about US $46,250 at the 2004 exchange rate) to house all former military personnel in Chiang Mai Province who had no offspring to support them. This retirement home initially had more than 130 residents, but that number had been reduced to about 30 in 2005 as many of them passed away over time.

The second program involved economic aid to improve general living conditions. In Maehong, for instance, the Taiwan government invested 780,000 baht (or US $19,500) in 1987 to build a water purification system in the hills above the village that would supply running water to the village houses. That same year, the Nationalist government also provided 1 million baht (US $25,000) to pave the village streets. The village government estimates that the total financial assistance to Maehong village alone amounted to 10 million baht (or US $2.5 million at the 2004 exchange rate) over the next dozen years.

The Nationalist government in Taiwan also provided technical assistance to Yunnan Chinese in northern Thailand. Agricultural extension workers and researchers were sent from Taiwan to promote new crop varieties and production skills. In addition, teams of handicraft masters were sent from

Taiwan to teach villagers how to make souvenirs for tourists. The impact of these forms of technical assistance seems to have been more significant in Chiang Rai Province where many Yunnan Chinese turned their hill farms into tea gardens and the export of tea has become an importance source of revenue. In Chiang Mai Province and Mae Hong Son Province, the agricultural teams taught villagers to grow tree crops such as plums and peaches, but with limited success. Many Maehong villagers complained that the types of fruit trees produced low yields of inferior quality, and they were soon abandoned.

The willingness of the Taiwan government to admit children of the Yunnan Chinese to Taiwan for college or professional education from the 1960s through the 1990s is a practice that has played crucial role in these communities' development, yet its actual monetary value is hard to calculate. Most of the children who completed degrees in Taiwan have stayed on to find jobs there. In Maehong village alone, an estimated 1,000 village children have gone through this educational channel and settled in Taiwan. Remittances from Maehong offspring working in Taiwan provide steady income for many villagers.

MAEHONG VILLAGE

The village of Maehong was named in relation to a Thai village immediately to Maehong's east. It was built immediately below the rugged mountains on the Burmese-Thai border where the Third Field Army has had their bases since 1962. Originally a small trading post occupied by local Thais and Karens and surrounded by virgin forest, this foothill community sits above the rice terraces that are the dominant agricultural feature of the Fang-Chiaprakarn basin, about 140 kilometers north of Chiang Mai City.

When American historian F. W. Mote and his wife visited this village in the mid-1960s, the area had just been turned over to the Third Field Army for their permanent residence, and the Thais and Karens no longer lived there. Mote reported a village population of 680 and predicted that since transnational immigration from China and Burma would be unlikely to continue in the foreseeable future, the village would not likely grow much. This prediction did not prove correct, and Maehong grew continuously over the next few decades, with continuous trickles of refugees from both war-torn Burma and politically volatile China, especially during the Cultural Revolution years (1966–76). Mote and his wife were, however, very perceptive in pointing out that this village had been very successful at agricultural production, and with its well-run village government and good Chinese education programs (Mote 1967).

When the village was built in 1964, few villagers had the personal savings to purchase farmland. Some of them labored for Thai paddy farmers in nearby villages for meager daily wages. Others ventured into nearby hills to clear forest slopes and grow seasonal crops. With their extensive knowledge about hill farming and keen interest in external commodity markets, villagers oriented their agricultural production to the demands of the fluctuating market and invested whatever profits they obtained in the purchase of additional farmland. Very quickly, within two decades, villagers owned not only large tracts of slope lands purchased from the slash-and-burn Lisu or Miao who retreated into the deep forest when faced with the encroaching Yunnan Chinese, but also the surrounding paddy fields they purchased from Thai farmers, their previous employers, in nearby villages.

Today the village population is close to 10,000 in about 1,100 households, and immigrants from China and Myanmar still cross the steep mountains to find refuge here. The village's former *phuyaiban* (headman)[6] estimated that about one-third of the current village residents are original settlers and their offspring who came in 1964. The second third includes middle-aged immigrants who settled in the village in the 1970s and 1980s, mainly escapees from China's turbulent Cultural Revolution and Burma's civil wars. The last third are new immigrants who arrived within the last 10 years; these include both Chinese who came to the region through Myanmar in search of a better life and Burmese hill tribes (Burmans, Lisus, Karens, and Shans) trying to escape Myanmar's ongoing civil conflicts (Feingold 2000).

Initial impressions indicate that this village is well organized, with rectangular residential blocks, paved streets, and well-built, well-maintained houses, a clean environment, well-lighted streets, and well-stocked retail stores along the main thoroughfares. It also appears to be peaceful and tranquil: old Yunnan Chinese soldiers greet one another in the street to chitchat about daily events, and children play freely in small groups, running up and down the streets.

Rough estimates made by the former *phuyaiban* indicate how prosperous this village has become: more than 75 percent of village families own automobiles (either sedan cars or light trucks), and 100 percent own motorcycles. Even middle school children go to school on motorcycles or mopeds. Maehong is the envy of the surrounding Thais and Karens, who have not only gradually sold or leased their lowland rice terraces to the villagers, but who have also turned to long- or short-term employment in either village factories or the orchards along the mountain slopes for their living.

ORGANIZATION OF THIS BOOK

To fully explore the dramatic changes we observed in Maehong and to provide a comprehensive understanding about their multi-layered implications, both pragmatic and theoretical, our research team conducted longitudinal and cross-sectional studies that combined diverse research methods that we hope will become a model for future research projects. A more detailed discussion of the research problems, the composition of the research team, and the methods used to gather qualitative and quantitative data is presented in the next chapter, chapter 2: "Posing the Research Problems." There I discuss the theoretical implications revolving around Chinese ethnicity and expand that discussion to include issues related to sociocultural sustainability. I also position Maehong in the emerging global community and illustrate how this case study may provide new insights on Chinese culture and society.

This is followed immediately by chapter 3: "Building A Sustainable Rural Livelihood," which provides in-depth analysis of the villagers' farm operation strategies and their level of agricultural sustainability. In this context I analyze factors contributing to the hill farms' seemingly successful operation. I also evaluate critically the concepts of sustainability and development when they involve other minority groups in northern Thailand's ethnic mosaic. Longitudinal quantitative data on changing soil and water quality have also been gathered to assess the sustainability of this agriculture regime.

Since the Nationalist refugees who settled in Maehong initially looked upon themselves as transient sojourners with the ultimate goal of returning to China, they were eager to preserve and reproduce their cultural traditions and ethnic identity. Chapter 4: "The Means to Cultural Reproduction: Education, Kinship, and Popular Rituals" focuses on how the Yunnan Chinese diasporas have preserved and reproduced their cultural traditions, including the establishment of the local Chinese school, the compiling of genealogy books, and the construction of the patrilineal extended family. Village founding fathers made tremendous sacrifices over the years to create this success and preserve this identity. What was produced in Maehong was a patriarchal power structures in the family and strictly gender-determined interpersonal relationships. For their offspring, however, this cultural heritage has more pragmatic implications than that of cultural preservation.

It is in this context I also discuss the major religious representations that we observe in Maehong. The popular rituals that the Yunnan Chinese brought with them to Thailand include organized worships, seasonal festivals and rites of passage. For the original settlers, they consider these ritual performances critical elements in cultural preservations. However, for their offspring, they seem to be quite willing to shed their cultural heritage which their forefathers arduously built for new social identity in the more tolerant Thai society. In this context, I analyze how this transplanted society, based on a reified Chinese moral order, has produced a petrified social entity in the northern frontiers of Thailand.

Chapter 5: "Emerging Internal Factionalism, Contested Political Loyalty, and Shifting Ethnic Identity," addresses issues related to internal tension and conflicts as the village enters the 21st century. Major problems derived from several sources. One is the declining power of the former military authorities. As the village has become more diversified, the former command hierarchy based on ranks and personal loyalty has become ineffective and irrelevant. The passing of General Li in 2000 signaled the end of the system of unquestioned patriarchal authority. Another factor is the Thai government's efforts to absorb these semi-autonomous borderland communities into its civilian administration system. The implementation of village-level elections, including the village mayor and village council representatives, paved the way for the rise of a new power elite who are genuine Thai citizens—not the naturalized Yunnan Chinese diasporas—and fully educated in Thai. This shift of village power has engendered fierce infighting and it may take a long time to resolve such conflict.

This chapter also engages in greater detail issues of changing ethnic identity among the villagers and their children. While our initial impressions indicate that the hill-farming system established by the original settlers has been both economically viable and ecologically sound, these farmers, like other Chinese immigrant communities in Thailand, face the possibility of being absorbed into mainstream Thai society. As a consequence, the avowed goals of many first-generation villagers, namely to preserve their cultural heritage and, more subtly, their Chinese ethnic identity, seems to have become more difficult to achieve. More detailed analysis, however, indicates that this identity shift among the younger generations may not be so out of ordinary after all. As border-crossing refugees, they may have already possessed the de-territorialized characteristic that positions them as the primordial transnational. In that sense, whether they maintain their Chinese identity or not become a moot issue.

The last chapter, Chapter 6: "Conclusions," summarizes the major points raised by this project and identifies possible areas for future research.

THE SIGNIFICANCE OF THE MAEHONG PROJECT

Even though many of the initial findings here are tentative, they nevertheless provide important heuristic insights into pressing social issues that may be relevant to other region or nation-states elsewhere in the world. In this light, social scientists and policy-makers may gain a useful understanding about the nature of these intricate problems and their possible solutions; this could be useful when dealing with similar issues in their own societies.

The first issue we tackle here is opium eradication and the identification of alternative cultigens to replace this form of livelihood. The failure of most campaigns in eradicating poppy production in many parts of Asia, especially in northeast Burma, has frequently been related to the failure to find suitable alternative cultivars for reclusive hill farmers who often reject crop replacement programs due to low financial return of the new crops or limited accessibility to external markets. The success of the Yunnan Chinese in northern Thailand with their profitable hill farming system has proven that exceptions do exist, and their practices deserve close scrutiny. The Maehong case is especially significant in that the refugees voluntarily weaned themselves from drug production, consumption, and trafficking without coercion from the outside.

The second issue we tackle is the nature of ethnicity and the possibility of building a pluralistic society devoid of group conflict derived from divisive social categories. The yearning of human beings to construct a utopian society that is genuinely free from prejudice based on arbitrary social categories (i.e., socioculturally constructed), be it religious, ethnic, class, gender, linguistic, and so on, remains an unfulfilled dream. The ideal of a pluralistic and multicultural society that fosters mutual respect across ethnic, racial, gender, and national boundaries, and that inculcates among all citizens an appreciation and respect for diverse livelihoods, is an elusive goal that only exists in academic discourse. It is in this context that our empirical study may provide useful insights.

The transition of the Chinese diasporas into mainstream Thai society raises several interesting issues. The first is the nature of Thai society. How have the Thais, being "a loosely structured society," as John Embree remarked (1969),[7] been able to accommodate the continuous influx of immigrant groups into their society with minimum conflict while maintaining their unique cultural heritage and identity? The successful absorption of Chinese into Thai society, often moving through specific trades or classes or professions, seems to imply a trade-off of ethnicity for selected social status. This observation further raises the question of what exactly does the concept of "ethnicity" mean? Does it matter at all in a relatively open and flexible pluralistic society—or an approximation of it—such as Thailand? In the same vein, as globalization has become an increasingly compelling social force that permeates nation-state boundaries and transnational a common feature in life, what lessons can we learn from the Yunnan Chinese and their host society, Thailand?

The third issue this project raises is the nature of "sustainability." While many scholars or policy-makers accept this concept simply as an axiom of best practices, our study argues that the matter is far more complicated than originally thought. The most critical problems revolve around the absence of objective criteria or measurements with which we can evaluate how sustainable a system is (Fricker 1998; Levin 1992; Solow 1992). Are we dealing with the issue of an ecosystem's "survival" potential based on human activities—hence a state of affairs that can be measured with input-output analyses? Or are we dealing with a process in which the involved parties consciously engage in deliberations about the future of humanity—a consciousness-raising process geared towards achieving an ideal? To make the concept of "sustainability" operational, we need to do much grounded analysis about how local discourses are incorporated into overall community deliberations when charting out community productive plans.

A related problem is the subjective perspective or standpoint from which we assess an ecosystem's sustainability. Development projects that produce a sustainable livelihood for one group may inadvertently have negative consequences for another. When we move from one level of analysis to another, the meaning and implications change. Researchers starting from different standpoints often arrive at dramatically opposed results. In other words, in discussing issues related to development or sustainability, we must ask what is the unit of analysis: the village that the anthropologist resides in, the geographic region that an agronomist analyses, the nation-state about which the macro-economist compiles data, or the whole biosphere that concerns the environmentalist? How we resolve the complex issue of differential scales—both temporal and spatial—when measuring sustainability deserves our careful considerations (Fricker 1998; Levin 1992; Solow1992). While I do not pretend to provide direct or fully satisfactory answers to all these questions or issues, it is my hope that the empirical findings of this case study will provide useful examples for future scholarly discussion.

This project also touches on policy dimensions related to hill farming, land ownership, and the legal rights of hill tribes in Thailand. To preserve the original forest coverage in the northern hills where most of the major rivers in Thailand originate, and to prevent flooding caused by soil erosion in the hills, Thailand's Royal Forestry Department has initiated programs to stop slash-and-burn agriculture and restrict the movements of hill tribes (Rerkasem et al., 1994; Santasombat 2003; Sturgeon 2005). The restricted residence of the migratory hill groups and the denial of their legal rights are considered two major issues among the hill tribes that have resulted in mismanagement problems according to empirical studies (Rerkasem et al., 1994; Santasombat

2003; Sturgeon 2005). The Yunnan Chinese hill farming system could be regarded as an alternative to these established policies.

Finally, this project is significant for the field of China studies. Researchers on Chinese culture and society have taken many theoretical positions about its nature and operational principles. Maehong represents a community transplanted *en masse* from its original sociocultural milieu and rebuilt in Thailand's remote jungles that is patterned on an idealized Confucian moral order; that order is characterized by graded interpersonal relationships and prescribes that social superiors bestow benevolence on their lessers and, conversely, that loyalty goes the opposite direction; the gendered power hierarchy favors male/men/father/husband over female/women/daughter/wives; ancestral veneration is expressed in written genealogy and funeral rites; and filial piety underscores the supremacy of family integration and continuity at all costs. The questions we may ask in this context are: Does this type of living arrangement best meet the needs of life in the contemporary global scene? How will their graded interpersonal relationships based on relatively stable long-term reciprocal rights and obligations be changed as these Chinese are increasingly engaging in transnational movements and activities? When this sociocultural complex is juxtaposed with the Thai social milieu, known for its de-emphasis on extended kinship organizations or multigenerational families, as well as for its relative gender equality (though not in the political sphere), we may expect to see changes among the Yunnan Chinese. To complicate this issue further, Chinese culture and society has been changing rapidly in mainland China, Hong Kong, and Taiwan in recent decades. When villagers from Maehong encounter their counterparts from the other Chinese communities, what type of cultural representations will they be adhering to? How might such encounters affect their sense of identity as Chinese and efforts at cultural preservation? Furthermore, since diasporic conservatism or anachronism is known in the literature, such as among Jews and Koreans in United States (Han 2005; Sarna 1999) or Pakistanis in England (Husband 1998), will the Yunnan Chinese efforts at cultural restoration be considered an anachronism? These questions will be explored in detail in the remaining chapters of this book.

NOTES

1. In the Chinese historical context, Mark Elvin's famous discussion of "the high-level equilibrium trap," which prevented China from entering the industrial revolution mode in the 14th and 15th centuries, is surely a historical predecessor (Elvin 1977).

2. Funding for this project was provided by the Wenner-Gren Foundation under its International Research Collaboration Programs in 2002–3 and 2004–5, and by Iowa State University's Bridge Fund (2003–4).

3. The scholars cited have different opinions about the terms "Ho," "Haw," or "Chin Ho," but most agree that they apply mainly to Muslim Yunnanese caravan traders and hence are unsuitable designations for the Nationalist soldiers. Forbes and Henley (1997), however, argue that these are generic terms used by residents in northern Thailand for all Chinese and hence should be used.

4. The author's follow-up book, *Jinsanjiao, bianqu, huangcheng* (金三角。邊區。荒城。) *The Golden Triangle*, *Border Lands,* and *Deserted City,* Bo 1982), created quite a lot of publicity in Taiwan and generated many assistance programs for the Yunnan Chinese. This book, however, was based on the author's short excursion to the region and hence does not have the same level of in-depth knowledge about local conditions.

5. Again, researchers have presented different estimates of the number of soldiers who retreated from Burma to Thailand during this period. Forbes and Henley (1997) estimated that the Third Army had about 1,500 soldiers while the Fifth Army had 2,000. I believe that Chang's numbers are more reliable (Chang 2001, 2002).

6. The Thai system of local autonomy for tribal areas by appointing a village chief, responsible for all administrative affairs, more details in Chapter 5.

7. Embree's original thesis, first published in 1950, has generated a lot of debates among researchers in Thai studies. Later works by Aphichat Chamratrithirong (1984) and Shin'ichi Shigetomi (1992), however, largely confirmed Embree's viewpoint, although they pointed out that the notion should take into consideration two modifying factors: first, there were ostensible regional differences, such as Central versus the Northeast Thailand, in their degree of looseness; and, second, the recent penetration of the market economy that might have affected the pre-modern reliance on dyadic relationships—the basis of loose structure—in Thai society.

Chapter 2

Posing the Research Problems

Two cross-cutting research interests brought me to Maehong, a Yunnan Chinese village in a remote corner of northern Thailand. My first research interest concerns the issue of sustainable agriculture or sustainable rural livelihood—a hot button issue in the heartland of the United States at the turn of the century. Being in one of the major land-grant universities in Iowa, the core of the American corn-belt, I was involved in organizing and teaching a new graduate program on agricultural sustainability and was keenly aware of the shortage of empirical studies based on grounded research. Ethnographic case studies, with their in-depth local knowledge and detailed database, can provide comparative frameworks to understand the actual operation of an "ideal" sustainable system and significantly enrich and expand the scope and theoretical issues in this field.

The second research interest I have in the current project is more personal. I have been conducting anthropological fieldwork in rural areas of China, Taiwan, and Hong Kong for over two decades, dealing with issues such as land reform, changing farm management and productivity and, more recently, children's nutrition and health. Knowledge gained from these diverse settings, although still within the larger Chinese cultural orbit, has provided heuristic insight to deepen my understanding about Chinese culture and society, such as its familial mode of division of labor, pre- and post-natal child care, labor exchange in farm work, gender hierarchy, community-based rituals, and so forth. I was ready to expand my ethnographic reach beyond the conventionally called "China proper" and to examine how the alteration of the external environment in which the people live might have changed their ingrained cultural practices. This expanding research interest led my casual reading about Chinese diasporic communities in northern Thailand to a more systematic inquiry and, ultimately, a research project. Let me elaborate on how these two cross-cutting interests, one institutional and the other personal, have merged together to form the current research project.

The call by American agricultural scientists to conduct research on sustainable agriculture at the end of the 20th century clearly indicated that their growing awareness about the potential short- and long-term ecological damage the U.S. commercial farming might have brought about was something too big to be ignored. The rising concern signaled that even scientists in the world's most advanced agricultural productive regime in the heartland of the U.S. had decided to take action, either preemptive or remedial, to rectify their own seemingly successful record. That was exactly what I thought in 2000 when I was invited to join a committee composed mostly of agronomists and soil scientists, to establish a graduate program on sustainable agriculture at Iowa State University.

The issue of agricultural sustainability was particularly relevant in the state of Iowa. Due largely to its successful deployment of modern scientific outputs, including industrial products (e.g., hybrid crop varieties, farm chemicals, combustion combines, etc.,) and innovative technological know-how (no-till farming, conservation tillage, remote sensing, GPS etc.,) this agrarian state had been enjoying nature's bounty in the form of high-yield crops (chiefly soybean and corn), their added-on values (i.e., feed lots for hogs, fowl, and cattle), and other related food processing industries, that had provided a livelihood envied by subsistence farmers around the world. And yet in spite of these ostensible successes, both farmers and research scientists have been aware of the non-sustainable features in agriculture that were popping up all over this idyllic landscape: the loss of farm topsoil due to wind and rain erosion; the pollution of natural streams and water reservoirs due to farm chemical runoff; the repugnant gaseous odors from animal feed lots or excessive agricultural chemicals that threaten community livability; the potentially disastrous short- and long-term health hazards for humans caused by pharmaceuticals used in animal feeds; the uncertain environmental and health impacts of genetically modified crops; the rising costs of industrial products in farm operation that cut into its profitability; the anticipated exhaustion of world petroleum reserves and other non-renewal resources that constitute the foundation of this agrarian regime; and, most alarmingly, the out-migration of rural youth who moved to both coasts of the U.S. and cast a long and dark shadow over the state's future.

If commercial farming is not a sustainable farming system to meet humanity's basic food quest, then what other alternatives do we have? Small-scale and labor-intensive farming with intermediary technology may minimize some of the most obvious environmental problems that American agribusiness has engendered. And yet, at a time of increasing global competition, how can such small-scale and non-industrialized farming systems compete successfully against the multinational agribusinesses and carve out specialized production niches to meet the basic conditions of sustainability? Even if they could identify locally specialized cropping systems and occupy the market niches, could they successfully reproduce the next generation of farmers to ensure community self-renewal?

It is in this context that I became particularly interested in the Yunnan Chinese in northern Thailand's Golden Triangle region. They have tried to establish a rural livelihood in the precarious tropical hills where any human intervention in the environment can easily tip the delicate balance and produce direct and irreversible damage. They have also switched from seasonal crop production (mostly vegetables and opium) to agro-forestry. Most importantly, the Yunnan Chinese diasporas have professed to preserve traditional Chinese culture and, implicitly, ethnic identity. Can they do better than Iowa farmers to keep the younger generations on farm? In short, here we have a small-scale community that can be examined closely to understand how its cultural components have been conducive to or, adversely, retarding the efforts toward the "ideal" state of sustainable development. A study such as the current one will not only shed light on issues related to agriculture sustainability, it will also provide new insight on Chinese culture and its change in new sociopolitical contexts.

After studying extant publications about the contorted history of the Yunnan Chinese diasporas in general and the complex interethnic relations in the Golden Triangle, I was convinced that this is a project worth pursuing, and formulated a funding proposal to carry out my fieldwork in Maehong. To assess the success or failure of the Chinese diasporas in their transition from opium production and trafficking to tropical fruit production, I realized that employing a multi-dimensional yet holistic approach that encompassed expertise from diverse academic disciplines, as well as combining qualitative and quantitative data collection procedures, offered significant advantages. It became apparent to me at this point that a research project of this magnitude could only be accomplished by a team of researchers who take up different aspects of the transition and yet are congenial enough to share basic data and research findings that would contribute to a comprehensive picture. Through numerous phone calls, email messages, and letters, I was able to recruit a multidisciplinary and multinational research team for the task. My successful application to the Wenner-Gren Foundation for Anthropological Research, under its International Collaborative Research Program, permitted us to launch the project in late 2002.

1. THE THREE RESEARCH PROBLEMS

The three main research foci of this project are: the ecological adaptation of the Yunnan Chinese hill farming system in Maehong and its vicinity, the shifting Chinese ethnic identity among the diasporas and their offspring, and the significance of the nation-state among the diasporas in the emerging global society. Let me elaborate on these three themes separately before I discuss the formation and division of labor amongst our team members.

Ecological Adaptation in Northern Thailand

Most of the current residents in Thailand's northern hills have been settled within living memory, except for small pockets of Mon-Khmer–speaking groups, such as Lua and Htin (Kunstadter 1983: 28; Li 2001a, 2001b; Young 1962). Their slash-and-burn agriculture primarily revolves around subsistence needs such as hill rice and corn, or cash crop production such as opium. Efforts to improve their living conditions, first initiated in 1965 by the Tribal Research Centre in Chiang Mai and later by other national and international development agencies, have included the introduction of paddy rice and wheat to replace indigenous crops and the resettlement of the hill tribes to lowland areas (Geddes 1983; Li 2001a, 2001b; Rerkasem et al. 1994; Santasombat 2003; Sturgeon 2005).

The arrival of the Yunnan Chinese and their introduction of vegetables and tropical fruits rapidly expanded the subsistence-based agriculture and the cash-cropping varieties in the area. Chinese financial management skills (Young 1962: 32) and ability to build ethnic marketing networks (Auansakul 1995: 33; Maxwell-Hill 1998: 98) also made viable the extension of local agricultural production to food processing, which meant that local crops could travel by market networks far beyond the hills. In recent years, the production and marketing of litchi, mangos, longan and honey tangerines in hill orchards, as well as other indigenous seasonal produce such as bamboo shoots, baby corn, and ginger root has fueled a boom in canning factories and trucking companies, for both domestic and international markets.

How will this newly introduced agricultural regime and its spin-off industries impact the local environment and its short- to long-term sustainability? More specifically, I want to understand: what technologies did they use to retain topsoil in the hills and to preserve soil fertility at the local level? Where did the Chinese receive information about new crop varieties, market conditions, or other modern industrial inputs? Will the application of modern farming ingredients, such as chemical fertilizer, pesticide, and herbicide, in hillside fruit orchards create soil and water pollution, and change the biosphere structure permanently? And, finally, are the hill farming knowledge and technology transferable across cultural boundaries so as to benefit other hill groups who try to extract themselves out of the dual scourges of poverty and drug involvement?

Practical knowledge from our project may be useful for development projects not only in highland northern Thailand but also in hill regions of other Southeast Asian countries and southwestern China. With our in-depth knowledge about local farmers' hill farming practices, we hope to acquire empirical data to assess their level of sustainability. In doing so, we also hope to isolate specific sociocultural factors that may be conducive or a hindrance to building a sustainable rural livelihood.

Shifting Chinese Ethnic Identity

As the Yunnan Chinese have become more prosperous in Thailand's northern hills, they face the same choices as other Chinese immigrants to Thailand: they can either maintain their cultural heritage and ethnic identity as sojourning Chinese, or they can participate more fully in Thai society with a new sort of identity (Bao 1995, 2005; Easman 1986; Keyes 1987; Maxwell-Hill 1983, 1998; Purcell 1951; Skinner 1957; Wang 1996; Wyatt 1982). There is a long history of Chinese immigration to Thailand, especially from the mid-19th through the mid-20th century, the time when Western colonial powers rapidly expanded their economic engagement in Southeast Asia and required a large number of craftsmen, laborers, middlemen, retailers, and so on, to cope with an ever-expanding trade (Easman 1986; Keyes 1987; Pongsapich 1995; Purcell 1951; Reid 1996; Skinner 1957, 1996; Wyatt 1982). Immigrants from China's southern and southeastern coasts quickly filled the various economic roles and managed to occupy pivotal positions in many trades. However, in contrast to Chinese diasporic communities in other colonial settings in Southeast Asia, the Chinese immigrants in Thailand have been quite willing to be assimilated and have maintained a relatively harmonious relationship to the host society.

Researchers such as Purcell (1951), Easman (1986), Keyes (1987), Wyatt (1982), and specifically, Skinner (1957, 1996) have offered an assimilation theory to explain the Chinese immigrants' easy passage into Thai society. First, unlike Chinese in the Dutch East Indies or British Malaya where they needed to learn the official language of the colonial masters but not the local languages, Chinese immigrants in Thailand learned only the official Thai and so could easily associate with the locals. Second, most early Chinese immigrants were single males and they appreciated Thai women's marketing skills and management ability. Inter-ethnic marriage was common. Many offspring of such unions found it easy to participate in Thai social life while maintaining their Chinese ethnicity. Third, the Thai royal family has long made a practice of bestowing honors or high status on successful immigrants, including Chinese, thus making assimilation attractive since it implies upward mobility. And finally, the similarities between the Thai and Chinese, especially in their cuisine, religious practices (e.g., both accepted variants of Buddhism, with the major difference in that the Thai practice the Theravada version while the Chinese incorporate some of the Mahayana beliefs into their popular rituals), and even folklore (Reynolds 1996), made this transition easy. Skinner's model thus postulates a linear, progressive transition that extends from "immigrant Chinese" to "Sino-Thai," and to "authentic Thai," a process that can be completed in two to three generations.

Many scholars have challenged this assimilation model. Instead of looking at a linear progression from "immigrant Chinese" to "Sino-Thai" and then to "Thai," Coughlin (1960) argued that the Chinese immigrants maintain a double identity, or "dualism," that permits them to shift from one category to another with ease. Similarly, Bao (2005), Chan and Tong (1993, 1995), and Maxwell-Hill (1998) argue that Chinese ethnicity in Thailand should be understood as a multilayered cultural construction that defies Skinner's progression-assimilation model.

We must point out immediately that the Yunnan Chinese in northern Thailand are significantly different from other early Chinese immigrants discussed by previous researchers. They are distinct in several important ways: region of origin (southwest China rather than south and southeast China); language spoken (Southwest Mandarin versus a variety of regional languages such as Cantonese, Teochiu, Hakka, and Holok from Hainan Island); location of settlement (northern Thailand versus central and southern Thailand); modes of migration (overland rather than by sea); length of time since migration (second half of the 20th century versus 19th and early 20th century); and the nature of the migration (the Yunnan Chinese were forced out of China and migrated involuntarily, while the early migrants entered Thailand voluntarily in search of economic opportunity). Finally, most of the early Chinese migrants concentrated in urban areas, especially Bangkok, whereas the Yunnan Chinese in the north are mostly rural, generally ignored by researchers because of their rural settings (Heidhues 1996).

The Yunnan Chinese are different from other Chinese immigrants in Thailand in yet another important area: their ideological commitment to anti-communism, which motivated them to escape China in the first place. As self-proclaimed carriers of authentic Chinese cultural heritage, the Yunnan Chinese have made strenuous efforts to maintain and perpetuate it: They live in nucleated—and hence more homogenous—villages and have clearly delineated military ranks and command structures, all of which make this cultural reproduction project feasible and more easily achieved. The Yunnan Chinese have also established more explicit family- or community-based projects (e.g., ancestral graves, Chinese school, Daoist temples, and other voluntary associations) to sustain their cultural traditions, and at the same time they construct more explicitly negative ethnic stereotypes about others and so, by enclaving, have maintained their distinct sense of ethnic identity longer than the other Chinese immigrant groups.

In spite of these differences, the Yunnan Chinese and earlier Chinese immigrant groups share a similar and fundamental set of core sociocultural values, social relations and practices, including patrilineal descent in inheritance and kinship reckoning, patrilocality in post-marital residence, a male-centered gender hierarchy, and filial piety and ancestral veneration—the pre-modern Chinese cultural traditions that have been lumped together and characterized as Confucianism for the sake of clarity (Bao 1995, 2005; Chan and Tong 1993; Keyes 1987; Maxwell-Hill 1998). Furthermore, the reliance on extended interpersonal social networks, be it regional (Skinner 1957) or professional (Coughlin 1960), also constitutes an important form of shared "social capital" (Lin 2000; Ong 1999) among all Chinese immigrant groups.

If, indeed, ethnic identity is both multilayered and situational as the scholars cited above have argued, we can still investigate how these layers are composed, prioritized, manipulated, and presented. For instance, we may ask: at what point would immigrants or their sons adopt the local practice to become Buddhist novices in a Theravadan temple? When will they adopt Thai surnames and give up their Chinese family names? How long will the Yunnan Chinese and their offspring carry on the traditional Chinese popular rituals—seasonal, spiritual, familial, and political—in constructing daily life? Similarly, even though there may not be a distinct social category of people who call themselves "Sino-Thai" and who represent the transitional stage between Chinese immigrants and mainstream Thai society, Amara Pongsapich suggests the emergence of a Sino-Thai mode of operation: "The third and/or fourth generation Chinese have adopted a Sino-Thai cultural pattern. Some still practice ancestral worship and visit the graveyards of the ancestors while others have abandoned the practice" (Amara Pongsapich 1995: 23; a similar note was also made in Keyes 1987). Since assimilation of Chinese into local societies in Southeast Asia has been well documented elsewhere (Reid 1996; Skinner 1996), it will be useful to identify the possible phases of change and the core versus peripheral cultural markers used in maintaining their ethnic identity.

The Nation-State in a Transnational World

If the twentieth century can be characterized as "the century of modernization," the twenty-first century is, without a doubt, the century of globalization (Appadurai 1996; Edelman and Haugerud 2004; Inda and Resaldo 2002; Lewellen 2002; Sassen 1996; 2007). In this new century we have witnessed the restructuring of the global social, political, economic, and cultural orders. The erasure of old and inscription of new spatial and temporal demarcations on existing ethnic and nation-state boundaries have engendered both conflicts—such as the renewal of long-dormant ethnic rivalry and animosity in the former Yugoslavia—and opportunities, as seen in the increased cooperation and population movements across ideological and geographic divides in the post-Cold War era. The emerging global market under the guiding principle of neoliberalism has challenged traditional notions of nation-state, state sovereignty, and the meaning of ethnicity (Edelman and Haugerud 2004; Inda and Resaldo 2002; Lewellen 2002; Sassen 1996; 2007). Correspondingly, these developments have heightened academic interest in the study of diasporas (or diasporic communities), transnationalism, and the emerging globalization itself.

Anthropologists have been particularly sensitive to the issue of globalization because the local communities in which they conduct research are often situated at the fringes of global flows of events. As Jonathan Inda and Renato Resaldo eloquently phrased it:

> The world today is witnessing an intensification of circuits of economic, political, cultural, and ecological interdependence. This is a world in which the radical acceleration in the flows of capital, people, goods, images, and ideologies—subjects and objects, in short—across the face of the globe has brought even the most remote parts of the world in contact with metropolitan centers. . . . It implies a fundamental reordering of time and space. (Inda and Resaldo 2002, 5).

These constantly changing spatio-temporal dimensionalities must have been acutely felt by the Yunnan Chinese in Maehong as they engage in transnational movements. We may ask how they fit into this changing global configuration as they assert their presence in Thailand. The China they left behind was a war-torn chaos, and yet it was idealized as a Confucian monolith with a more or less stable social order, as well as shared ethical standards and moral principles. Their border-crossing into Burma and later into Thailand was part and parcel of an effort to preserve this idealized cultural heritage. To this day, the first generation settlers believe that their diasporic communities in northern Thailand are temporary and transitory, to be discarded when the global anti-communist tide turns in their favor and they can reclaim cultural orthodoxy for the motherland.

The global sociopolitical developments of the second half of the 20th century, however, haven't borne out those expectations. The cultural China they brought in their mind no longer exists—definitely not in the Mainland, nor in Taiwan, Hong Kong, or other overseas Chinese communities. The end of the Cold War did in fact signal the global retreat of communism as an ideology, but it also rendered the Yunnan Chinese ever more marginal in the postmodern world—not only geographically, but also as a petrified diasporic community in the midst of a pluralistic Thai society.

Most scholars acknowledge that the presence of diasporic communities has contributed to the resurgence of ethnic, religious, and nationalistic aspiration and tensions (Ong 1999; Sheffer 1986; Van Hear 1998; Weiner 1993) and hence increases the potential for political instability and conflict, especially between the immigrants and their host societies (Demmers 2007; Husband 1998; Stubbs 1999). Nevertheless they also point out potential benefits these communities have introduced into their new environment. The first is the formation of a multicultural society: the presence of ethnic diversity has been shown to increase the possibility of the host society's self-renewal and cultural regeneration. The second is the ability of diasporic communities to form networks and channels that transcend territorial or national boundaries. Both Easman (1986) and Ong (1999) credit the success of Chinese diasporas in Southeast Asia primarily to this ability and the resulting enhanced fluidity of both capital and information. Living in a more flexible and tolerant society (Embree 1969; see also Thongchai 1994: 4), how have the Yunnan Chinese facilitated cross-regional trade, as well as the flow of labor, capital, and information?

In this study, we will be especially interested in the human actors in various global settings, because their stories best suggest how the Yunnan Chinese have positioned themselves in a diverse social environment with equal levels of empathy, involvement, and effectiveness in communication. Maxwell-Hill attributes their successful adapting to their multicultural orientation: "The Yunnanese lived in the midst of cultural and linguistic diversity in their native province, and northern Thailand is not much different in this respect" (1983: 131). Similarly, referring to the success of overseas Chinese in Southeast Asia in general, Ong argues that the "Chinese entrepreneurs are not merely engaged in profit making, they are also acquiring a range of symbolic capital that will facilitate their positioning, economic negotiation, and cultural acceptance in different geographic sites" (1999: 17–18).

In this study, I want to explore the institutional mechanisms the Yunnan Chinese use to establish their global reach. Questions that can be asked include: Could they represent a sort of primordial transnational entity who voluntarily moved across the nation-state boundaries in pursuit of a better life and adapted to the new global social order by strategically deploying their tangible and symbolic capital, including both the social and the cultural ones? Besides language, ethnicity, and marriage, factors already identified as key by anthropologists who study China, what other types of interpersonal *guanxi* (networking) can be used for this purpose? Conversely, if ethnicity and nation-state citizenship are nothing more than instruments that can be used and then shed in this global arena, where will the Yunnan Chinese anchor their identity in this amorphous world? As the globalization process accelerates, what new personal identity schemes might be developed and deployed to meet the deep-seated need for belonging and attachment? From where will the Yunnan Chinese draw the inspiration that defines the successes and failures of their present life and afterlife?

2. RESEARCH METHODOLOGY

Since only limited ethnographic fieldwork has been conducted among the Yunnan Chinese diasporas,[1] providing a systematic account of the people and their daily life will be an essential first step. To assess the success or failure of their efforts at building a sustainable community and reproducing Chinese diasporas, we must begin by establishing a set of baseline data and then measure its changes over time.

A number of research methods used in anthropology, agronomy, and geography were employed by participating researchers to gather data on the various facets of the project. The data that I draw upon for this study mostly involved a qualitative approach—recording in detail the villagers' farming practices (including hill farming techniques, fruit production and processing, labor management, and marketing), major social events, and aspects of daily social life. Through informant interviews I investigated how the villagers operate their hill farms, market their produce, select future spouses, secure education for their children, and participate in popular rituals.

To come to an understanding of the hill farming system, I asked a number of questions: How did they select the original slopes for farming? Where did they acquire the knowledge about crop varieties and planting? What modifications or innovations did they make so that introduced cultigens survived in this new environment? How is information about cropping technology spread after successful experiments? Changing weather and vegetation coverage data gathered through GIS were also systematically collected to help us assess the speed and degree of expansion of the hill farms.

To assess the environmental impact of these farms, we obtained soil and water samples from selected sites over the research period to assess their changing physical and chemical profiles. The longitudinal and cross-sectional comparative data thus obtained should clearly pinpoint the positive as well as negative features of the hill farm system.

In addition, my participant-observation of village life and interviews with key informants interrogated the daily life of the villagers to come to an understanding of their predisposition to multicultural living. Four areas were of particular interest: first was the formal efforts by which villagers invest in cultural reproduction—their establishment of a Chinese school and the construction of community facilities that answer their spiritual needs (e.g., village temples and cemeteries) and other communal facilities. The enormously important Chinese education provides useful channels through which the village youth can achieve upward mobility in one of two ways: Either they obtain a college education and acquire professional status in Taiwan or by virtue of their multicultural education they gain advantages over other Thai youth who, with their mono-cultural education, are less equipped to move into better positions in the increasingly global Thai labor market. Thus, the Chinese education at once manifests and delivers a reified ethnic identity that helps the local Chinese link up with other Chinese and so gives them broader access to the possibility of social mobility than ordinary Thais might have. To study the specific roles and operations of the Chinese education, we interviewed teachers, students, school administrators and, whenever possible, schoolchildren's parents.

The second area of interest is kinship reckoning and recognition, spousal selection, and post-marital residence—all based on reified Confucian practices. Does the traditional practice of patrilineal kinship reckoning and inheritance still prevail? What about patrilocality in post-marital residence? Have changes occurred with villagers of the younger generation who have been exposed to the alternative Thai model of living where kinship relationships beyond the nuclear family are of lesser importance? What about the traditional Chinese gender hierarchy that gives males/husbands/fathers unchallenged authority over females/wives/daughters? We gathered this data primarily from interviews. Additional data such as genealogy and ancestral tablets will also be photocopied when available.

The third area of interest centers on symbols and rituals. The performance of rituals associated with popular folk beliefs in family contexts, in specifically built localities or sites, and on official ceremonial occasions can be clear indicators of cultural reproduction and a shifting personal ethnic identity. Are traditional Chinese seasonal rituals still performed? What types of cultural symbols are used for weddings, funerals, and other ceremonial occasions? How are Thai cultural symbols or seasonal rituals incorporated into local praxis—consciously and unconsciously? What do the Chinese think of the Thai monks and temples they sometimes employ for ceremonial occasions? At what point will the parents send their sons to Thai temples as novices, following Theravadan version of Buddhism, and abandon their selected Mahayanan practices? To get at these issues we relied on direct observation, videotaping, and informant interviews to compile the divergent opinions on such practices.

Finally, I am concerned with social organizations. Do the Yunnan Chinese participate in (Chinese) regional associations, or professional organizations? Are official positions in these organizations regarded as symbols of status and ethnic marking? How are interpersonal relationships cultivated through gift-giving and banquets? How does the community officially interact with external authorities such as the Thai civil administration or the military? And how do they engage with visitors from Taiwan and China—officially, privately, and through charitable and commercial organizations? Our interviews with organization officials and, when possible, access to their historical records were helpful in shedding light on these issues.

3. FORMATION OF THE RESEARCH TEAM

During the proposal-writing period, I assembled a team of researchers who share similar interests and agreed to take part in this collaborative project. The composition of the team is both multidisciplinary and multinational, and our diverse interests complement each other in more than one way. I will discuss the six participants' personal expertise and interests first.[2]

Professor Pong-In Rakariyatham, Professor of Geography and Vice President of Chiang Mai University, used Geographical Information System (GIS) technology to reconstruct geomorphological change in the vicinity of Maehong village. Changing distributions of hill farms versus the secondary forest can be demonstrated through the retrieved data over the past fifty years (i.e., 1954, 1978, 1983, 1986, and 1995).

Associate Professor Mattiga Panomtarinichigul, a soil scientist at Chiang Mai University, has long been involved in assessing the impact of hill farming in northern Thailand, and the possibility for improvement through experimentations with new cropping systems. In this project, she identified five hill land use patterns: litchi orchards, mango orchards, tangerine orchards, mixed crops farms, and secondary forest in Maehong and its vicinity. She and her graduate assistants established fixed schedules to do soil measurements and take soil samples in order to determine the degree of soil erosion and the changing physical and chemical contents of both the soil and nearby streams that derived from chemical fertilizer, pesticides, and herbicides.

Professor Shu-min Huang, anthropologist, formerly Iowa at State University and now at Academia Sinica in Taiwan, focused on new crop varieties brought over to Maehong and its vicinity by the Yunnan Chinese, and the changes in cropping technology due to environmental constraints. He explored the significance of indigenous agricultural knowledge in hill farm adaptations, and the actual benefits of these new cropping systems in the local economy.

Professor Hui Yang, an anthropologist at Yunnan University, emphasized women's changing roles as the Yunnan Chinese moved across international boundaries and settled in Maehong. She investigated how women performed the role of keepers of historical memories and cultural tradition in this immigrant society, and as beholders of family integrity and continuity. The contrast between Chinese and Thai gender roles revealed significant differences about marriage, family, and meanings of life.

Mr. Jianyu He, a photographer and PhD candidate in Anthropology at Yunnan University, focused on ordinary family life, seasonal festivals/rituals, and life-crisis ceremonies (such as weddings, funerals, and graduations). The villagers' involvement in these activities indicates important cultural continuity and/or changes. Their participation in Thai versus Chinese ritual festivals signaled their shifting orientation in one or the other direction. Ethnographic films can often convey much richer visual messages than written expressions do.

Mr. Yin Duan, an anthropology graduate student from the Chinese University of Hong Kong, concentrated on the impact of Chinese education on the second and third generation immigrants in Maehong. He wanted to find out how the transmission of traditional Chinese cultural values affected the young people's ethnic identity and ethnic network maintenance.

In general, all four anthropologists emphasized long-term, in-depth community involvements through participant-observation. They established an intimate rapport with villagers and walked at ease to visit individual families or take part in community festivals. As a contrast, the geographer and the soil scientist conducted short-term site visits, garnered longitudinal data that are clear and precise, and identified major trends or problem areas that the anthropologists might be unaware of. The complementary and mutually beneficial relationship between these diverse approaches was quite obvious from the very beginning of this collaborative project. For instance, the anthropologists provided background information and history about different farming practices and about farm estates that the geographer and soil scientist generally had no knowledge about. In the meantime, before the geographer and soil scientists provided solid empirical data, anthropologists can only speculate the long-term impact of the hill farming system, especially the changing dynamics between the human agents and their environment. One thing is clear through this collaborative effort: while we gained in-depth information about the Yunnan Chinese and the village they built, we also gained new understanding about the other disciplines. This mutually beneficial experience enhances our mutual respect.

Due to our different institutional calendars and available time, we did not require all participants to visit the village at the same time, although we did manage to have several researchers stay in the village at the same time so that we may directly exchange our information and ideas. In late 2002, Huang, Pong-In, Yang, Duan, and He spent several days in the village at the same time. In summer 2003, Yang, Huang, and Duan all visited the village with overlapping schedules for about three weeks. During that time, Mattiga also spent a few days in the village. In summer 2004, Huang, Mattiga, and Duan visited the village with overlapping stays of about two weeks. Because of the requirement for fixed scheduled soil and water sampling, Mattiga made the most frequent visits to the village. Due to personal reasons, Yang and He from Yunnan University dropped out of the program entirely after 2004.

A second year Wenner-Gren Foundation grant permitted the project participants to organize a panel presentation at the 2004 American Anthropological Association's annual meeting, originally scheduled in San Francisco but later changed to University of California at Berkeley. Initially, all six participants agreed to take part in this panel, with five papers (by Huang, Pong-In, Mattiga, Yang, and Duan) and an ethnographic film (by He). However, because of internal problems at Yunnan University, Professor Yang and Mr. He decided not to attend at the last minute. The turmoil at the 2004 AAA meeting further complicated the venue of our presentations. Luckily, through an arrangement with the Society of East Asian Anthropologists, our panel met in Berkeley as originally scheduled. These four papers provide useful data for the current ethnography.

NOTES

1. The two exceptions are Dr. Wen-ching Chang, who has worked among the Yunnan Chinese since the mid-1990s, and Dr. Jian Li, who worked with northern Thailand's Miao population during approximately the same time. For more information about their work, please see bibliography.

2. All researchers were involved in this project during the 2002–2005 funding period. After the completion of their personal projects, however, they ceased to continue their involvement owing to personal reasons. I am therefore solely responsible for the final write-up of the project.

Chapter 3

Building a Sustainable Rural Livelihood

To properly interrogate whether and how the Yunnan Chinese in Maehong have established a sustainable agriculture or a sustainable rural livelihood in the tropical forest in Thailand's northern hills, we must first identify several interrelated issues embedded in the concepts "sustainability," "sustainable agriculture," and "sustainable development." In our assessment of the hill farming practices in Maehong, we must keep these problems in mind in order to gain a comprehensive picture of their successes and failures.

First is the nature of swidden agriculture, which has been practiced by indigenous hill tribesmen for generations. Scholars have debated whether this is a viable and environmentally friendly form of agriculture for the tropical rainforest (Fox 2001; Hansen 1994; Reed 1990; Sturgeon 2005; Young 1998). While most anthropologists support swidden agriculture as a practical way to maintain tropical agro-ecosystems and biodiversity, as well as a critical element in the survival of marginal tribal cultures (Anderson 1993; Bates 2001; Fox 2001; Geertz 1963; Santasombat 2003; Sturgeon 2005; Young 1998), others disagree, pointing out its negative effects such as soil erosion and the destruction of vegetation, as well as its poor quality as a management regimen for natural resources (Knox 1989; Rerkasem et al. 1994; Wright 1993).

Relatedly, what might the parameters of "sustainability" be? The concepts of sustainable agriculture, sustainable rural livelihood, or sustainable development have broad appeal to a wide spectrum of audiences, but there seems to be little consensus about what might constitute the necessary and objective criteria by which to measure the success or failure of a system (Francis 1990; Fricker 1998; Gold 1999; Hatfield and Keeney 1994; Helmore and Singh 2001; Roling and Wagemakers 2000; Santasombat 2003).

Finally, there is the question about the viability of any sort of farming in tropical rainforests. Alarming reports about the disappearing rainforests and dwindling biodiversity in Madagascar, for example, are representative of the sharp criticisms of farming in this ecosystem:

> With 90 percent of the forest gone, uncounted species have lost their habitats and become extinct, and most of the soil cover has been lost to erosion . . .
>
> This destruction is largely due to slash-and-burn agriculture. People are constantly burning the rain forests for agricultural purposes—it is the way they make a living (Wright 1993: 451).

If farming in tropical rainforests is to be permitted, it must come through consideration of the conflicting demands of environmental preservation and biodiversity, population pressures and their demand for farmland, long- and short-term economic development planning, and concern about the worsening global greenhouse effects (Anderson 1993; Fox 2001; Rerkasem et al. 1994; Santasombat 2003; Young 1998). So who should have a say in deciding these policies—the land-hungry farmers, the environmental conservationists, both local and global, researchers of various stripes, or the state's economic development officials?

These issues defy simple answers and present no uniform criteria for objective assessment; their consideration is, furthermore, inevitably informed by the subjective perspectives researchers bring to the field: our disciplinary convictions about social problems and proper solutions, our personal value judgments when evaluating practical actions, and our moral commitments to certain ethical tenets and perceived justice. Hence the research protocol of our current project takes an additional step beyond ordinary ethnographic fieldwork: instead of observing or evaluating the actions and performance of our research subjects, the Maehong villagers, we begin by reflecting on our own positions, motivations, and value judgments in designing and carrying out our project. Are we advocates for the interests of the marginal and/or marginalized peoples in the hills? Are we promoters of economic development of specific ideological stripes, such as neo-liberal, conservative, or conservationist? Are we environmentalists who have a vested interest in preserving biodiversity and in soil and water conservation? And, finally, are we implementers of government policies regarding land and forest management? At this level of self-reflection, we may disassociate ourselves temporarily from these issues and consider the subjective implications of concepts such as *sustainable agriculture, sustainable rural livelihood* or *development.*

Let's use one of the simpler definitions of sustainable agriculture to illustrate the delicate nature of the problem. As defined by John E. Ikerd, "A sustainable agriculture is one that is capable of providing everlasting value to society. A sustainable agriculture, so defined, must be ecologically sound, economically viable, and socially just" (Ikerd 1992). In other words, the chief concern of our study is the 'society'—an aggregate of people (i.e., the Yunnan Chinese) with certain shared cultural traits and identity, and our goal is to determine how they (members of the society) realize the "everlasting value" of environmental soundness, adequate material provisioning for life sustenance, and justice for all involved parties.

Herein lie the conceptual difficulties revolving the issue of sustainability: whatever constitutes 'sound,' 'viable,' and 'just' are subjective terms with explicit value judgments. They are based on the needs and assessments of the very group that we are interested in. Are we (the researchers) the decision makers who determine the parameters of our studies and the degrees to which their needs have been satisfied? Furthermore, a sustainable system suitable for one group may be built at the expense of other groups, and one that suits a particular time span may become non-sustainable at another time. In other words, the contents and meanings of sustainability may change when we move from one level of analysis to another or from one locality to another. When we take into consideration different scales in terms of space, time, and local social complexity (Levin 1992; Solow 1992), the whole concept becomes difficult to define. With this awareness in our study of the hill farming practices in Maehong, we must keep the above-mentioned problems in mind in order to gain a comprehensive picture about their successes or failures.

CHANGES IN HILL FARMING PRACTICES IN MAEHONG

When Maehong was established in 1964, the current village site was very much covered by the original rainforest. Small patches of rice paddies owned by Thai farmers in nearby villages sat to the east. In the hills to the west lay a few patches of clear-cut forestland owned by the swidden Lisu tribesmen. Villagers recall today that life was hard then because the wages they earned from the Thai paddy farmers were low, and the labor input for clearing the slopes was high—they made hardly enough to meet basic subsistence needs.

Their first break came in the late 1960s, as several of my informants recalled. When the Vietnam War intensified, refugees from Laos and Cambodia left their homes and entered northeastern and eastern Thailand where refugee camps were established along the Thai border. Many international donor agencies charged with feeding the people in these camps soon discovered that Chinese cabbage was a relatively durable and inexpensive vegetable the refugees found acceptable. The demand for cabbage drove up prices, and many Maehong villagers turned their hill slopes into cabbage fields for quick profits. It was during this time that several entrepreneurially oriented village farmers began to act as purchasing agents who collected cabbage for the camps and hooked up with external vegetable market dealers, mostly *Teochiu* (潮州, *Chaozhou* in Mandarin) Chinese, from cities such as Fang, Chiang Mai, and even Bangkok.

Connections with other overseas Chinese communities, especially the powerful Teochiu people, turned out to be critical to Maehong's future development. In the early 1970s, two of Maehong's most enterprising farmers, based on information provided by Teochiu wholesalers, turned much of the hill farmland to growing potatoes, not only in Maehong but also in nearby Yunnan Chinese villages under the control of the Third Field Army. They also served as purchasing agents or wholesalers who acquired potatoes from individual farmers and delivered them to the external markets. Increased demand for potatoes came mainly from two sources; one was the expanding U.S. military presence in Vietnam and the second was the rising middle class in Thailand, which consumed potatoes in the form of snacks, such as potato chips.

Early on the villagers' agricultural practices were highly extractive in nature. Seasonal crops such as cabbage and potatoes exposed the tilled ground directly to the sun. During the monsoon season, soil erosion was significant as runoff water from the hills carried away heavy sediments that turned the streams reddish. During the dry season, hardly any vegetation grew on the exposed soil and the landscape around the village turned barren. Many informants today acknowledge that the area was not a pleasant sight during those years, yet they had little incentive to search out alternative crops that might be both profitable and environmentally benign. At that point, they were not legal residents of Thailand and had no legal rights over the land, so they couldn't have cared less about the long-term damage their practices might cause. Had the status quo continued, the Maehong villagers and the effects of their livelihood on the ecosystem were so negative to render it unsustainable.

The watershed event that led to a dramatic change in this human-nature dialectic came in the early 1980s. After the successful military campaign of 1981, in which the Yunnan Chinese helped the Thai military wipe out all insurgent Miao communists in the region, the Thai government granted citizenship to the Yunnan Chinese military personnel and their families. Gaining Thai citizenship triggered a completely new *modus operandi* among the Maehong villagers, since as naturalized citizens, they now had the legal right to own land. They proceeded to purchase a few mountain slopes from the neighboring Lisu, Karen, and Thais, who considered hills to be of low agricultural value. Conscious efforts were made to increase land productivity through investment in farming technology as well as by experimenting with new cultivars. Equally important, the villages now made real efforts to maintain the quality of their newly acquired property through refined land management and farming.

The dramatic changes can be seen in the Table 3.1, drawn up from GIS sources:

Table 3.1. Changing Land Use Patterns in Maehong, 1954–1995

	1954	*1978*	*1983*	*1986*	*1995*
Village size (in acres)	0	92	125	209	237
Crop land, un-irrigated	214	754	364	286	502
Paddy, irrigated	82	373	363	355	121
Orchard	0	81	218	265	566
Forest	1644	640	870	770	514
Total	1940	1940	1940	1885	1940

Adapted from Rakariyatham, 2004.

Several trends can be identified in this table. The most obvious are the gradual increase of village size, the dramatic increase of the orchard lands, and the steady reduction of the natural forest. The fluctuation of un-irrigated cropland size needs explanation. The jump to 754 acres in 1978 is a clear reflection of the villagers' effort to expand their production of cabbage and potatoes. The reduction in 1983 and 1986 signals the reduction of seasonal crop production in favor of cultivating fruit trees. The same pattern can be seen for the paddy fields. Increasing rice production, originally considered a desirable goal, and hence a focus of significant investment until 1978, has also gradually given way to orchards.

Changing land-use patterns in Maehong and its vicinity tell only half of the story, however. The new hill farming system must also be assessed in terms of its sustainability. If we define a *sustainable rural livelihood* as a holistic, systematic, and participatory approach that brings together issues of poverty reduction, along with environmental and social sustainability, the measure of this approach must be the people's ability "to effectively address needs for food and income, cope with and recover from shocks and stresses, and maintain and enhance its capabilities and assets while not undermining the natural resource base" (ISU Sustainable Rural Livelihood Website; see also Helmore and Singh 2001).

For the Yunnan Chinese in Maehong, this means their mode of farming must combine considerations of environmental conservation, improvement in living conditions, and sociocultural renewal. To accomplish these goals, they had to decide which crops to experiment with and from what sources.

They also needed to acquire the technologies that would preserve soil fertility in order to maintain the long-term productivity that shifting cultivation does not provide. In the monsoon region of northern Thailand, the torrential rains from May through November mean that proper conservation methods should be employed to contain soil erosion. Conversely, the four-month-long dry season from December through April requires sufficient irrigation water to ensure plant health and growth. Additionally, to carry out all these tasks, farm operators must hire long- or short-term farmhands to supplement family labor. In short, achieving a sustainable agriculture or sustainable rural livelihood requires proper agricultural knowledge, adequate investment capital for technology acquisitions, a risk-taking capability to experiment with new crops, and the managerial capacity to organize sufficient farm labor for the tasks. The following sections will lay out the particular adaptations that led to the seemingly sustainable livelihood in Maehong.

Yunnan Chinese Agricultural Expertise

The Yunnan Chinese brought with them a number of new cultivars and cropping technologies when they first came to this region. Before fleeing China, they had mostly been farmers in the hilly western section of Yunnan Province. They were familiar with intensive farming in river valleys and basins where rice is grown in paddy fields, as well as with hillside farming using terraces. They came armed with rich cultural knowledge about crop varieties and hill cultivation.

The Chinese introduction of vegetables and new tropical fruit varieties has expanded the local subsistence-based agriculture and cash-cropping far beyond what the swidden agriculturalists had produced in the past. As Gordon Young observed early on:

> They (i.e., the Yunnan Chinese) . . . demonstrate a type of mountain living, which has never before been seen among Thailand's hill tribes. Their example is important, even should they not remain, in terms of what might be shown the other hill tribes in future. They had brought ploughshares from Yunnan and use these to advantage in their agricultural practices, using oxen to pull them. In addition, they have shown the importance of buckwheat as a staple and as livestock feed. Their vigor in animal husbandry has also been a source of great admiration by the Lahu and Lisu tribesmen living around them (1962: 84).

Villagers in Maehong claim that when they first settled here, the neighboring Thai and Karen farmers only grew paddy rice and tended some fruit trees such as mangoes and longan. The Yunnan Chinese brought with them new vegetables, such as cabbage, Chinese lettuce, potatoes, water kale, turnips, eggplants, string beans, and various gourds. Chinese lettuce (called *woju* 萵苣), for instance, is a crop that the Thais had never seen before.[1] Villagers also claim that neighboring Thai farmers did not know about grafting as a technique for growing trees. Only after working for the Yunnan Chinese for several years did they learn this technique.

Besides crop varieties, the Yunnan Chinese have also been very knowledgeable about other aspects of agricultural production. They pointed out the necessity of using chemical fertilizer as an essential ingredient in modern agricultural productivity; at the same time, they also acknowledged that continuous use of chemical fertilizer may harden the topsoil and lower soil quality. Organic nutrients are still needed, and the villagers use cow dung purchased in bags from Chiang Mai for this purpose in the fields; they apply it intermittently in orchards.

To ensure an adequate water supply during the drought season, villagers searched the limestone hills behind the village for natural springs. On locating them, they built cement storage tanks around the water sources at high elevations. Using gravitation, they channel the water through plastic pipes to the slopes below. This availability of water meant that the villagers could rapidly expand their hill farm cultivation.

The Introduction of Tree Crops

New crop varieties and animal domestication aside, Chinese financial management skills (Young 1962: 32) and the ability to build ethnic marketing networks (Auansakul 1995: 33; Maxwell-Hill 1998: 98) have also provided an essential ingredient for the development of a cash crop economy in the hills. In recent years, litchi and tangerine have become the two most successful crops, and their introduction was made possible thanks to networking between the Yunnan Chinese and other ethnic Chinese.

Litchi (荔枝) was introduced by Maehong villagers in the early 1980s, although it is barely mentioned in Edward Anderson's authoritative work on plants in northern Thailand (1993). Villagers claimed that they acquired litchi saplings from the Teochiu people of eastern Guangdong Province, the largest Chinese immigrant group in Thailand, who brought the seeds from southern China and established litchi orchards in the low hills around Wanyang township, about 30 kilometers to the northeast of Maehong, in the 1970s.

The success of the Teochiu merchants at establishing litchi orchards in Wanyang, which earned them sizable profits, prompted the Maehong villagers to follow suit. In 1981 two of the village's most skilled farmers, Mr. Yue and Mr. Liu, began to plant litchi on their hill slopes. They purchased most of their saplings from a Teochiu merchant who operated a large trucking company in Wanyang. With 20 litchi plants per *rai* (a Thai unit of about 40 × 40 meters), Mr. Yue planted 400 saplings on 20 *rai* and Mr. Liu set out 3,000 plants on his 150 *rai*.

There are four major litchi varieties: three are early ripening (*Jingzhong* 金鐘, *Japaka* 糯米糍, and *Ohe* 藕黑), and the fourth (*Fenghua* 豐花), is late ripening. Mr. Liu recalled that whenever the Teochiu merchant developed new varieties, he would give free saplings for Liu to experiment with. The merchant also used his trucking business to transport the harvested litchi to markets in Chiang Mai and Bangkok.

The honey tangerine (蜜橘) was brought to Maehong in 1998 in the same manner as the litchi was in the 1980s, that is, it was first introduced by Teochiu Chinese in Wanyang who had acquired seeds from southern China and successfully raised tangerine orchards in northern Thailand. The Yunnan Chinese acquired the seeds and saplings from the Teochiu people (tangerines were not mentioned in Anderson's book) and successfully established them in the surrounding hills.

Honey tangerines are even more capital intensive than the litchi. While each *rai* of land can grow 20 litchi trees, it sustains up to 80 smaller tangerine trees, increasing investment up front. While litchi orchards require only monthly irrigation during the drought seasons, the tangerine fields require weekly watering, especially during the dry season. Litchi orchards require fertilization twice a year, once before flowering and again immediately after the fruits emerge. But tangerines need monthly fertilization with both chemical fertilizer and cow dung. In addition, the tangerine trees need more regular trimming, grass removal, and pest control than litchi. In other words, while one *rai* of litchi orchard requires an investment of about 20,000 baht per year, the same size tangerine field takes about 250,000 baht, or greater than 10 times more.

Litchi and tangerine have different growth rates. Litchi can be harvested after the fifth year, initially producing small yields. The tree matures and reaches its peak at about 12 years. It is not unusual to see a full-grown litchi tree produces 400–500 kilos of fruit each year. Villagers claim that its quality declines after the tree reaches 25 years. In contrast, tangerine trees can yield fruit around the second year of growth, and reach maturity in the fifth year.

In recent years, many Maehong villagers have replaced their litchi trees with tangerines. They gave three reasons for this change. First, there is an overproduction of litchi in Thailand now. With rapid production growth in other parts of northern Thailand, market prices for litchi have been greatly depressed since 2000. Most villagers claim they haven't been able to break even with their litchi crop in recent years. Second, tangerine production can be better manipulated to correspond with the market demand and hence command better prices. The tangerine flowers in early January and this lasts for 45 days. After that, the fruit will grow for 10 months until harvest time in November, December, and January. Many villagers deliberately postpone tangerine growth by picking off the first flowers so the flowering will be pushed back to February. The tangerines will then mature one month after the regular peak season and command better prices. Villagers claim that there is no comparable technique for manipulating the litchi's growth and harvest. Finally, villagers pointed out that waxed tangerines hold up for ten days to two weeks longer than unwaxed. This longer shelf-life permits the fruit to be shipped to distant markets. Ripened litchi, on the other hand can be kept for no more than a week, so market region for them is necessarily smaller.

The rapid expansion of honey tangerine orchards in Maehong and its vicinity has worried many villagers. They fear that the continuation of this expansion could lead to overproduction as with litchi a few years ago, and hence declining profit margins. Preemptive strategies to deal with this problem had already been discussed by several villagers in summer 2004. One young entrepreneur named Huang proposed that a factory be built to process the fruit and turn it into tangerine juice concentrate or canned fruit for export. Another proposal I heard that summer, advocated by at least two villagers, Mr. Mei and Mr. Peng, was to introduce Sunkist oranges from Taiwan to gradually replace the honey tangerines. Mr. Peng's daughter-in-law, who originally came from Taiwan, actually brought five kilos of Sunkist orange seeds back to Maehong that summer and asked me whether grafting is needed when the seeds turned into saplings.

To accommodate to the increased output of honey tangerines, Mr. Huang, the former village *phuyaiban* and owner of the village's canning factory, began an export business to move fresh tangerines to Bangkok, Hong Kong, Singapore, and southern China in the summer of 2005. He told me that he would add two new lines in his factory in the near future. One would be a waxing facility to process fresh tangerines for export. The second would be a juice concentrate-making facility to process tangerine juice. This kind of foresight indicates that the Yunnan Chinese are hardly subsistence farmers, but rather farmer-entrepreneurs and investors who aim to maximize and maintain profits.

Investment and Conservation Practices

When the Yunnan Chinese first settled in the Maehong area in 1964, small numbers of Thai and Karen rice farmers lived in the foothills and Lisu swidden agriculturalists worked the surrounding slopes. Several villagers, once they acquired Thai citizenship or legal residence in 1981, purchased land from Lisu and the neighboring Thai and Karen. For example, old Mr. Peng purchased 120 *rai* of land for 20,000 baht from a Lisu hill farmer after he had been granted legal residency in 1982. He now claims that he could sell the land for 1,200,000 baht, even though he has no formal legal title for the land. The expansion of the Maehong hill farms has caused most Lisu to move farther into the hills away from the Chinese settlers.

Before the Thai government began to strictly enforce restrictions on clearing forestland in the early 1990s, villagers could simply select mountain slopes for planting and chop down the original vegetation to establish temporary usufruct rights. Such rights are recognized by other villagers and can be transacted and bequeathed in the same manner as any other commercial commodity or property.

Mr. Yue, a skilled farmer, says that selecting the best slopes for farming involves several elements. First, the gradient must not be too steep to allow cultivation and human movement. Another essential condition is the availability of water for irrigation. But even if the slope is not too steep and if water is available, the land still may not be suitable for agriculture if it faces north. A northern exposure means insufficient sunshine, especially in the winter months, and hence poor harvests. Similarly, mountain slopes facing west are considered better than those facing east. East-facing slopes receive sunshine before noon—but the radiation and heat in those hours is insufficient, resulting in poor crops. Needless to say, a southern exposure is the ideal for farm plots in general.

In the early 1990s the Thai government, through its Royal Forestry Department, began to enforce the ban on developing new farmland in the national forest. All existing hill farms were measured and recorded, but the Forestry Department did not grant land titles to the existing owners (Santasombat 2003; Sturgeon 2005). This has created high anxiety among many villagers who had bought and continue to work hill slopes. Untitled land in the hills is worth about 10,000 baht per *rai,* or only about one-tenth of the value of farmland in the foothills for which legal documents exist; these can be sold at 100,000 baht per *rai.* In Mr. Peng's case, for instance, since he has no official title for his 120 rai hill slope purchased in 1982, he estimates its worth at 1,200,000 baht. With formal land rights, the land value could increase tenfold to 12,000,000 baht.

Profits realized from fruit trees in the hills have dramatically increased land values in Maehong and its vicinity. Villagers have not only expanded fruit production on the slopes, but have also purchased lowland paddy fields from Thai farmers and converted them into fruit orchards. Mr. Huang, a young village businessman who made a fortune from the wholesale potato trade and a potato chip business and who had discussions with me in the summer of 2004 about building a tangerine juice concentration and canning factory in the village, purchased slightly over 100 *rai* of farmland in 2003, mostly in the lowland area around the village, and hired over 40 Burmese refugees to run his tangerine orchards. With an initial capital investment of over 100,000 baht per *rai* for the land and another 20,000 baht per *rai* for the saplings and the installation of the sprinkler system, the total invested amount of 12,000,000 baht (or about US$300,000) made Mr. Huang's orchard operation no low-cost peasant enterprise.

As the investment in fruit production has increased, conservation practices have become critical to preserving this valuable property. One common practice the Yunnan Chinese have adopted is building terraces along the mountain slopes to enhance crop growth and prevent erosion. Several farmers have also built cement channels to guide runoff to lower ground or nearby brooks without cutting into the slope surface. Grass and underbrush around fruit trees are regularly cut but not removed. If grass and underbrush grow too high, they can harbor field rats that not only eat the fruit but also damage the root systems. Rats are also controlled through the use of poisons or trapping. Regular trimming of grass and underbrush allows the root systems to hold on to the topsoil and prevent erosion. When herbicides became available, village farmers experimented with them to see if they could effectively control or slow the growth of the grass and underbrush while saving human labor to trim them, and, in doing so, still preserve the topsoil by not destroying its root systems. This practice is widely accepted by villagers.

Another common conservation practice is digging shallow basins around fruit trees to hold in place fertilizer and water whenever these were applied, a practice that drew marveling comments from Professor Mattiga Panomtarinichigul, who professed to have never seen such methods used by any other hill groups. This way, fertilizers applied around the fruit tree, whether chemical fertilizer or cow dung, are not easily washed away from the root area by surface runoff. Newly developed hill farming practices such as these, which make farming more environmentally sustainable, are becoming more common throughout the hills of northern Thailand.

Labor and Hill Farm Management

The Yunnan Chinese have had an even greater impact in this region with their farm business management skills and the use of credit. They are extremely careful in calculating investments, expenditures, and market prices for their produce, and reckoning the profitability of their work.

Investment in litchi and tangerine orchards both require large sums of capital for the initial purchase of land, tree saplings, and fertilizer; for the installation of irrigation facilities; and for labor costs. Because litchi take five years to start producing fruit and tangerines two years, this means that villagers growing these crops have to wait several seasons with no income from these investments.

Another important ingredient in hill farming is labor input. As many Maehong villagers grow old and their children migrate to Taiwan or to large cities in Thailand, they must find replacement laborers to fill this void. The intermittent political turbulence in northeast Burma over the past four decades has created a seemingly inexhaustible supply of refugees that have so far met the Maehong farmers' needs (Feingold 2000). Steady streams of Burmese hill tribesmen, including Wa, Shan, Karen, Lahu, and Mon have crossed the border to take refuge in Thailand. Given its close proximity to the Thai-Burmese border, Maehong is often the first stop of these refugees.

Maehong villagers possessed extensive knowledge about various minority groups and hill tribes even before they exited China, as Ann Maxwell-Hill eloquently points out: "The Yunnanese lived in the midst of cultural and linguistic diversity in their native province, and northern Thailand is not much different in this respect" (1983: 131). They have used this knowledge as the basis for hiring farm hands. In general, they preferred to hire what they call *baiyi* (擺夷) people. The term *baiyi* is without doubt a derogatory one applied generally to the Dai (傣) in Yunnan Province, or the Shan (撣) in northeast Burma, both distant cousins of the Thais.

Villagers claim that *baiyi* generally have a mild temperament and are not as feisty as the Burmans or Wa. They work hard and learn fast. Village farmers taught their *baiyi* farmhands basic skills for mixing and applying chemical pesticides and herbicides, the timing of irrigation and the trimming of fruit trees, how to detect and diagnose plant diseases; they have come to entrust responsibility for the entire farm operation to them, with just occasional supervision.

Since many *baiyi* farmhands enter Thailand illegally, they cannot live in the village as legal residents. Village farmers have built simple houses or sheds for them in the hills and provide them with a monthly wage of about 3,000 baht (US$75). It is very common for employers to hold this money for their farmhands and give them the accumulated sum when the employees want to return home or to visit their families across the border.

Villagers claimed that the best *baiyi* farmhands are young couples. They willingly work hard to save up money or help their kin on the other side of the border. Having a wife makes the male laborer less restive. Furthermore, if the male worker spends too much money drinking or gambling, the employer needs only to tell the wife about this problem and it will be fixed very quickly. Employer-employee relationships are often long-lasting, and *baiyi* farmhands are sometimes treated like family members. Whenever the Thai government has announced amnesty programs for illegal immigrants, employers apply for legal status for their farmhands to secure their long-term service.

The official record in Chaiprakarn Amphor indicates that 1,788 Burmese refugees (974 males and 814 females) resided in the vicinity of Maehong in July 2004. I believe that this figure is relatively reliable, although I do not know exactly what percentage of them were *baiyi* as opposed to the other ethnic groups such as Burmans or Wa.

Assessment of Environmental Sustainability

An objective assessment of the Yunnan Chinese hill farming system in terms of its environmental sustainability has been pursued through systematic field studies. Under the direction of Professor Mattiga Panomtaranichagul, a longitudinal study was carried out from August 2003 to July 2004. Five major types of land use were identified in the vicinity of Maehong: (i) maize, *Zea mays;* (ii) litchi, *litchi chinensis Sonn;* (iii) tangerine, *Citrus reticulate;* (iv) mango, *Mangifera indica Linn;* and, (v) secondary forest. Five catchments, each representing one of the above types, were selected for sampling.

Analysis of the physical and chemical properties of the soil was carried out on soil samples taken in the following periods: August 8–11, 2003;

Figure 3.1.

October 22 to November 24, 2003; February 21–24, 2004; and June 10–24, 2004. The soil's physical properties were measured at 0–20 and 20–40 cm depths in every sampling period. To ensure that the sampling was truly representative, data was collected from both upper and lower tiers of the slopes.[2] As mentioned above, the Yunnan Chinese make small basins around the fruit trees for the purpose of conserving water and maintaining soil fertility. To find out the extent to which this practice affects sustainability, samples were taken from both inside (I) and outside (O) these basins.

The water content profiles were also measured four times during the study period. At each site the survey team selected 5 to 8 locations to take soil samples using a soil auger for 5 depth ranges, at 0–20 cm intervals from 0 to 100 cm. The total amount of stored water in the soil was calculated at equivalent depth (mm) per 100 cm. Water content in forest soil was measured twice in February and June of 2004. Furthermore, samples of water, weeds, and stream sediments were taken three times for toxic substance analysis during the period from November 2003 through June 2004. The water qualities measured were nitrogen, phosphorus, water hardness, bio-chemical oxygen demand, acidity, and salinity.

The technical data and the analyses of this longitudinal study are too complicated and too technical to be present in total here (see Panomtarinichigul and Huang 2007). It will suffice to use some of the aggregate data to point out four general trends we identified that can shed light on our understanding of the practicality and sustainability of Maehong's hill farming system. First, it is clear that all types of agricultural land use, including the hill farming practices of the Yunnan Chinese, have predictably caused some degree of degradation, both of the physical and chemical qualities of the soil, when compared with that of the secondary forest. Forest soil has the best properties, including the highest values of organic matter, stable aggregates, infiltration rates, water storage, and so on, compared to all four types of hill farming. In other words, if there were no population pressures to open up the hill slopes for agricultural use, all efforts should be made to preserve the original forest (see Figure 3.2).

Second, when comparing the four hill farming regimes, it is clear that the fruit trees are more sustainable than the seasonal crops. In fact, the older an orchard, the better its soil profile. In Maehong, where most litchi orchards have been in existence for over 20 years, compared to the five or six years for mangoes and tangerines, the litchi orchards tend to have the deepest soil profile with the best physical and chemical properties. Maize cultivation results in the shallowest soil profile and the poorest soil quality. The wisdom here is, if population pressures or farmland scarcity forces farmers to cultivate hill slopes in Thailand's northern hills, governmental or development agencies

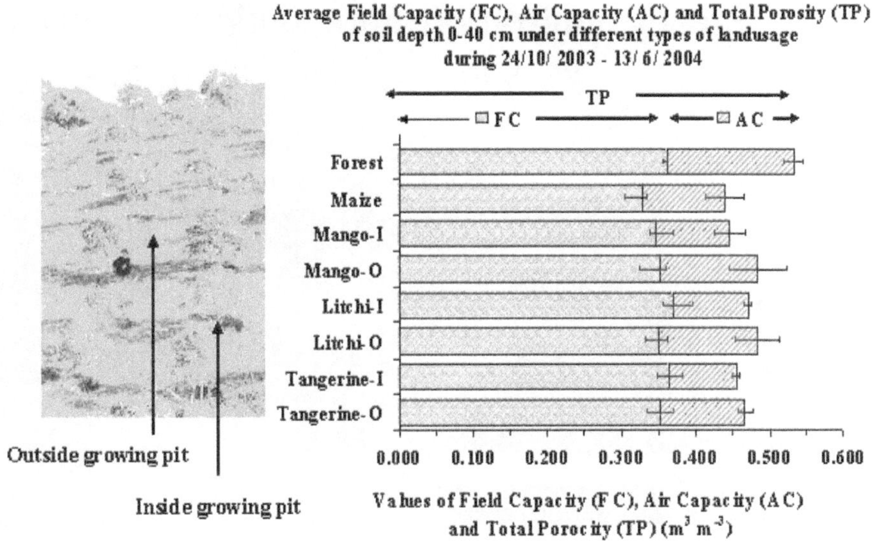

Figure 3.2.

should encourage them to grow perennial plants because they are more sustainable and environmentally beneficial than seasonal crops.

A third conclusion drawn from our study specifically addresses the cultural knowledge that Yunnan Chinese employ in hill farming. Construction of the basins around fruit trees is a practical soil and water conservation strategy. This practice, however, involves removing the top soil layer from the growing pits, and when this is done without mulching, it can result in poor soil quality with over-weeding and over-supply of chemical fertilizer and irrigation water. Comparing soil quality inside and outside the plant basins seen in Figure 3.2 and Figure 3.3. (e.g., Litchi-I versus Litchi–O or Mango-I versus Mango-O) consistently shows the discrepancy favoring soil outside the pit. In other words, these conventional practices deployed by Yunnan Chinese may actually lead to poor structure in the surface soil, resulting to a low infiltration rate and poor aeration, an imbalance of plant nutrients, and a consequent decrease in sustainable productivity.

The fourth and the final conclusion in our study indicates there is some degree of water contamination in streams from the over-use of chemical fertilizers and pesticides. Improper use of commercially available chemical inputs may produce residues of nitrogen, phosphorus, organophosphate, arsenic, and lead that find their ways into nearby streams and brooks during the rainy season. Even though the levels of most contaminants (except for lead) were still below the residual limit for drinking water, the presence of high levels of

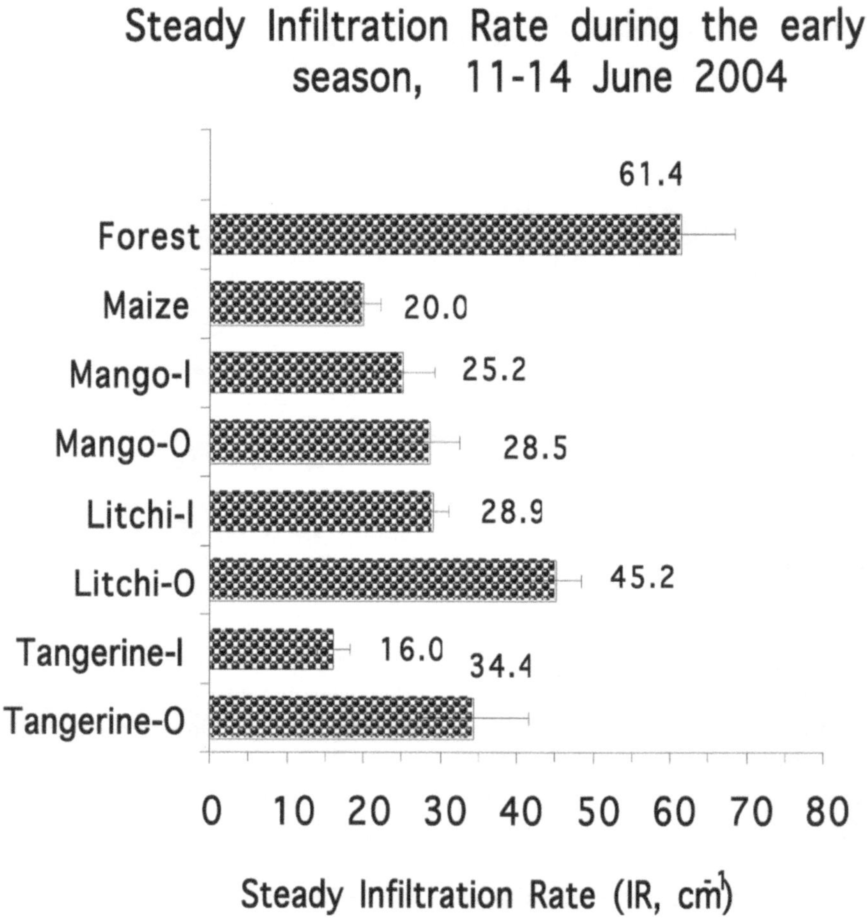

Figure 3.3.

lead was a clear signal that, unless properly remedied, the village's drinking water is already becoming a real health hazard (see table below).

SIDELINE ACTIVITIES IN MAEHONG

Besides farming, many male villagers also engage in sideline activities. Some young men have purchased vans or minivans with bank loans and use their vehicles to transport schoolchildren or hill produce. Even while working their farms, many village males (and some females too, although their number is much smaller) keep their eyes open for other possible employment

Table 3.2. Toxic Substances in Weeds and Water

		*Toxic substances in weeds**			*Toxic heavy elements in sediment***		
Sampling Date	*Sampling Location*	*Organo-Phosphate mg kg*$^{-1}$	*Organo-Chlorine mg kg*$^{-1}$	*Pyrethroid mg kg*$^{-1}$	*Arsenic, As µg kg*$^{-1}$	*Lead, Pb mg kg*$^{-1}$	*Mercury, Hg µg kg*$^{-1}$
24-Nov-03	Water drains at Lower slope (Site 1)	0.000	0.000	0.000			
21-Feb-04	Stream sideway (Site 1)	0.000	0.000	0.000			
11-June-04	Waterways in Litchi orchard (1)	0.000	0.000	0.000			
	Stream sideway (Site 2)	**0.010**	0.000	0.000			
	Lower stream line (Site 3)	0.000	0.000	0.000			
	Stream bottom and sideway (Site 1)	0.000	0.000	0.000	0.65	16.16	0
	Standard values for food and drinking water	**0.05–0.100**	**0.05–0.10**	**<0.050**	**0.05**	**0.01**	**0.005**

Note: The average amounts of the measured toxic substances and toxic elements (heavy metals), accumulated in weeds, stream sediments along the stream line started at the foot hill of litchi orchard, during August – November, 2003. in Maehong, Chaiprakarn district, Chiang Mai province, northern Thailand.

* Plant samples were taken from the stream side-wall and water drain channel.

** Sediment samples were taken from the bottom of stream line and water drain channel.

opportunities in and around the village. During the harvest seasons of the various hill crops, such as bamboo shoots, ginger, litchi, longan, and potatoes, villagers purchase quantities from the farmers and ship them to nearby wholesalers or factories for profit. Public construction work, whenever available, also attracts a large number of village men.

Many village women engage in food-processing work for wages. During the harvest seasons of litchi (late May, June, and early July) and longan (late July and August), village women congregate in a neighbor's courtyard, their children in tow, to peel and pit the fruit. The work is arranged in the following manner. A village woman who has a large number of female kin (e.g., daughters, daughters-in-law, husband's sisters or his brothers' spouses, and so on) or good friends, and who has a large courtyard to accommodate the assembled workers and their children, may serve as the contractor to do consignment work for a local factory. The fruit (either litchi or longan), in plastic crates, is collected by this contractor or her husband (or son) in the early morning around five from the local factory in a truck and delivered to the waiting women in her courtyard. Each woman worker brings a large plastic or tin basin for her finished fruit and a small wooden stool for sitting. Once the basin is full to capacity, it will be weighed and recorded by the contractor, and the peeled and pitted fruit will be stored in the original plastic crates. The processed fruit is transported back to the factory by the contractor in the evening for canning or drying. During work hours, the contractor will provide drinking water for the workers and toilet facilities for them when nature calls, and keep an eye on their children so that they would not run off to the street or cause harm. Another task the contractor must perform is to clean up the shells and pits after the work is done at the end of the day.

In 2006 village women earned 5 baht (or 12.5 cents) for each kilogram of litchi or longan they peeled and pitted. Since a skilled woman can handle about 30 kilos per day, she earns about 150 baht (or US$3.85). At the end of each day, the women receive slips recording the number of kilos finished and the sum owed for that amount. They take the slips at the end of the season to the contractor, who then pays them with cash received from the factory.

In 2006, the general condition in Maehong was that the contractor received six baht for each kilo of the processed fruit from the factory. This means that after paying five baht per kilo to the woman worker, she received one baht for each kilo of processed fruit that came from her consignment. If she had 10 workers that day, and each worker processed 30 kilos, she earned 300 baht, twice the amount of an average worker. There were incidental expenses that she must deduct from her income, however, such as gasoline for the truck or car, the fuel and water for the drinking water, and her own labor to supervise all the work operations.

The extensive use of credit in Maehong has had a significant impact beyond the village. Many nearby Thai villagers now bring their produce, such as longan or ginger root, to Maehong for sale, where they typically receive a promissory note from either the village food canning company or a village middleman. These transactions have not only tied Thai farmers into the global consumer market, they have also introduced them to the use of credit. Bank credit is commonly used in Maehong to purchase farm equipment and/or household goods such as an automobile, and this practice has been widely picked up by Thai farmers who maintain close contact with the Yunnan Chinese.

CONCLUSION

The Chinese diasporas in Maehong have established a proper rural livelihood, characterized by a hill farming system that is relatively environmentally friendly and economically sound. Much of their work has ridden on the cultural knowledge—agricultural, financial, and management-related—they brought with them when they entered the Golden Triangle. Their pecuniary acumen in saving and investment and their skills in building networks with other Chinese immigrants and in interacting with other hill groups, such as the *baiyi* people, have also contributed greatly to their operation. But the success of the Chinese has been built at the expense of the other groups—the Thais, Lisus, Karen, Shan—and in this context, both the winners and the losers must be equally assessed to determine the sustainability of this specific regime. Surely the issue of sustainability will have very different implications for the different groups and time span, and deserves our close scrutiny.

Can this hill farming system serve as a model for development, to be imitated by other hill groups to improve their livelihood? In its technical aspects this hill farming regimen can be easily transferred and acquired—as demonstrated by the skill of Maehong's *baiyi* farmhands. But in its other aspects, such as the long-term planning and investment, and the willingness to experiment with new crop varieties and technologies, this farming system demands another level of cultural habituation that may not be so readily adopted. The other hill tribes may not have sufficient investment capital as compared with the Yunnan Chinese when building their hill farms. Similarly, the critically important ethnic networking deployed by the Yunnan Chinese with other Chinese immigrant groups in pursuing technology transfer may simply not be an easy option for other hill tribes.

Yet another aspect of this hill farming system that may be hard to duplicate is the availability of refugees from Burma who continue to provide cheap labor for farm operations. The success of the Yunnan Chinese in Maehong has been to some extent built upon their ability to exploit other disadvantaged groups. Neo-liberal economists may argue that any contractual relations between two parties are established through mutual agreements that benefit both, and so without these employment opportunities, the Burmese refugees would probably face even greater difficulties in making ends meet. Thus the question of sustainability must return to the issues raised at the beginning of this chapter: what is our level of analysis and from whose perspectives? Are we dealing with just the Yunnan Chinese in Maehong, or are we dealing with the ethnic mosaic that includes not only the Chinese, but also other groups who interact with them on a daily basis for their livelihood? When we use the term "sustainable rural livelihood," are we referring to the perspective of the Yunnan Chinese, the Shan, or Thai development officials?

A second issue that requires consideration, based on findings from this project, is the current restriction placed on using forest for hill farming by the Thai Royal Forestry Department. Improving the living standards of the hill tribes has been a major concern of the Thai government. The twin goals of eliminating both opium production and poverty from the Golden Triangle can be best achieved by providing alternative modes of living, and replacing tropical rainforest with tree crops has been considered a viable and sustainable alternative (Fox 2001; Rerkasem, et al. 1994; Santasombat 2003; Young 1998). The income generated by the Yunnan Chinese hill orchards can be significantly expanded if current restrictions are modified. Once legal title is established to the productive orchard land, the Thai government can collect land taxes and property transaction taxes to augment its treasury. At the same time, granting legal title to the land should encourage increased investment in the hill farms. The locomotive effect of productive hill farms can not only remedy unemployment among the impoverished hill tribes, but may also sustain high production values for cash crops and processed produce that enter the international market.

Finally, at the epistemological level, we may question whether a sustainable agriculture and livelihood is a workable development goal for rural areas. These concepts have been developed over the past two decades as a countermovement to the dominant capital-intensive and industrialized agribusiness in the West; they have taken root among the proponents of organic farming, environmental conservation, intermediary technologies, and community food self-sufficiency. Incorporating environmental preservation into the industrial agriculture balance sheet has attracted tremendous scholarly interest (Francis 1990; Gold 1999; Hatfield and Keeney 1994; Helmore and Singh 2001; Ikerd 1992; Rerkasem, *et al.* 1994; Santasombat 2003; Young 1998). Our study indicates, however, that there are tremendous difficulties involved—both semantic and practical—when applying these concepts in real life situations. The effectiveness of the Yunnan Chinese hill farms changes when the scales of our analytic unit, both temporal and spatial, shift from one level to another. So do the perspectives from which we consider their practicality. This suggests that there remain a lot of problems to be teased out if we hope to make agricultural sustainability a viable option for the future.

The same problem may be related to the notion of "sustainable development," a concept that has contributed in an important way to new research directions in Western academies and to alternative lifestyles for rural practitioners in Europe and North America since the late 1990s, and yet it falls within a goal-oriented social blueprint for "development"—itself a concept that remains highly contested in this postmodern and postcolonial era.

Academic challenges to the dominance of development as a blueprint for societal change derive from two diverse sources. On one hand is the argument that developmental projects, especially those under the auspices of world capitalist hegemony, are not a solution but rather the cause of continuing third world poverty (Escobar 1994; Ferguson 1994a; Harper 2002). They argue that capitalist donor agencies such as the World Bank or USAID significantly contribute to perennial world hunger. On the other hand, the view espoused by James Scott argues that agricultural development projects orchestrated by modernist-oriented states, *both* capitalist and socialist, attempt to destroy natural complexity, cultural diversity, and local resistance, all elements abhorrent to the hegemonic nation-state, which is only interested in the issue of legibility (Scott 1998).

A third view that somewhat falls between these two positions is the characterization that external agencies engage in "organized intervention in collective affairs according to a standard of improvement," although both the notions of improvement and intervention may "vary according to class, culture, historical context, and relations of power" (Pieterse 2001: 3). It is by this moderate approach to development that Pieterse determines that, even in the poorest of the world's poor countries, the general conditions of health and education have improved in the last century (Pieterse 2001).

Development is a value-laden concept. To deny the poor around the world an opportunity to improve their less-than-desirable livelihoods because of those value connotations is commensurate to throwing the baby out with the bath water. A counter question then is: if we do not want to condemn the poor to such conditions permanently, what acceptable methods, generated either internally or externally, can be used to improve conditions, either in terms of living standards or sustainability? If we are to embrace the possibility that certain production methods may lead to improved material living conditions and alleviate hardship in daily life while preserving the natural environment and sociocultural autonomy and dignity, we may cautiously agree that the Yunnan Chinese mode of living in the Golden Triangle is a close approximation of what researchers suggest is a sustainable rural livelihood.

NOTES

1. Today, the Thais still call it *woju,* a direct borrowing of the Chinese term.
2. For more details, please see Panomtarinichigul and Huang, 2007.

Chapter 4

The Means to Cultural Reproduction: Education, Kinship, and Popular Rituals

Most Yunnan Chinese who escaped from China around 1949 were driven by a variant of anti-communist ideology: correctly or incorrectly, they perceived communism as a radical imported political doctrine that advocates class conflict and social division, and promotes the destruction of traditional Han Chinese culture, particularly a version of the reified Confucian ethics promoted by the Nationalists. Confucianism, a form of secular humanism primarily concerned with inculcating proper comportment between people to achieve social harmony, underscores the graded yet reciprocal interpersonal relationships that extend from the most basic values of filial piety (*xiao,* 孝) and benevolence (*ci,* 慈) between parent and child to other social domains. This familial model of normative behavior occupied a central place in pre-modern Han Chinese social life for centuries (Huang 2003).

William de Bary, a leading authority on Chinese philosophy since the middle of the 20th century, points out that Confucianism, as the state-sanctioned official ideology in pre-modern China, rested on three major pillars: the state administrative apparatus, formal education, and the family (de Bary 1988). He argues that when China entered the 20th century, the weakening of the imperial bureaucracy in the face of modern governance based on specialized technocrats and the replacement of traditional education with modern Western science-based education both diminished Confucianism's importance as an ideology. Confucian ethics have been gradually marginalized and relegated to only the familial domain (de Bary 1988). The importance of familism and its associated normative standards, such as male-centered kinship principles, family-based calculation for allocating economic resources, a gender hierarchy that marginalizes the female persona, and ritual practices that reify male virtues and female sacrifices, have acquired new ethical meanings among the Han Chinese and occupied much of the ethnographic record of the second half of the 20th century (Chuang 1985; Harrell 1985; Li 1985, 1986; Nee 1984).

While the decline of Confucianism in the Marxist mainland and in an increasingly globalized Taiwan reflect changing value orientations among Chinese intellectual elites since the late 20th century, we find that in Maehong traditional social ethics with heavy Confucian undertones still carry the day. Traditional Chinese values, such as filial piety, family cohesion, the supremacy of male heirs, loyalty to superiors, and unquestioned allegiance to one's Chinese ethnicity have been carefully cultivated through the a variety of public and private practices—family teaching, formal schooling, popular rituals, and so on, and permeate almost every aspect of the villagers' daily life.

In Maehong daily life, patrilineal principles of intergeneration inheritance and kinship reckoning still reign supreme: children inherit their family name and property from the paternal side of the family, and paternal kin are considered more important than maternal kin. The unchallenged domestic authority exercised by the patriarchal male is not only seen within the family, it is also expressed in the military hierarchy, as anthropologist Wen-ching Chang vividly observed in Maehong in the mid-1990s: the late General Li Wen-huan was treated as the grand patriarch of the Third Field Army: "some former soldiers even knelt down and kowtowed to him, an act which is normally performed to pay respect to one's parents, grandparents, and ancestors" (Chang 2002: 142).

Besides patriarchal authority exercised within the familial context, women's marginal position in the traditional social hierarchy is seen in their post-marriage residence—they go to live with their husbands and not vice versa, as well as their role serving their husbands' parents and making personal sacrifices to preserve familial integration and continuity. This unequal gender hierarchy forces all manner of suffering upon many women, while their men can indulge in gambling, drug addiction, taking concubines, and other irresponsible behavior with impunity.

Another manifestation of this discriminatory gender division is seen in marriage patterns. Hypergamy for women is an entrenched traditional Chinese practice—for men are supposed to be superior to women, and not vice versa. Interestingly, this value orientation has a significant impact on the villagers' ethnic identity and marital practice: Poor Yunnan Chinese males who have little or no means to marry Chinese women do not hesitate to marry tribal women who are considered inferior to the Chinese in general; still, their offspring are universally accepted as legitimately Chinese. This point will be covered further in chapter 5.

To preserve and reproduce a reified Confucianism via conscious adherence to traditional Han Chinese moral tenets, Maehong villagers participate in three distinct sets of practices that dominate almost every aspect of their daily life. These are the support and delivery of Chinese education programs, following social mores in which familism reigns supreme, and participating in public rituals—both seasonal and life-crisis ceremonies. This chapter describes how these practices are inscribed on villagers' lived experience.

BUILDING THE CHINESE SCHOOL

Yunnan Chinese are generally proud of their commitment to Chinese education, whether in Yunnan or in settlements abroad as in Burma and Thailand. Chinese language instructions, using texts that emphasize the Confucian virtues (filial piety and proper graded social relations) are regarded as the most important means to preserve Chinese cultural values. Villagers from Kogan in northeast Burma never hesitate to point out that villages in that area with substantial Chinese diasporas always have at least one Chinese school, and their pupils' knowledge of the Chinese classics is, in general, far superior to those graduated from comparable schools in either Yunnan or Thailand.

The same high esteem for a classical education was also seen among the Nationalist military contingents when they settled in the Golden Triangle. Maehong villagers say that whenever the troops stationed in a locality for a few days, the unit commander would immediately ask one or two of the better-educated officers or soldiers to give instruction on the Chinese classics to the military children or dependents who traveled with the troops. Formal education is still considered the most important means of cultural reproduction.

In 1965, the year after Maehong village was established, General Li Wen-huan ordered the construction of Zhongxing Elementary School (中興初級小學). At the outset, the school had one teacher and twenty-seven students. Mr. Duan, the first graduate of the school, recalls that when the school was first built, villagers cleared the forest coverage and used the felled timbers to build simple wooden shacks for classrooms and for the teacher's living quarters. With the rapid inflow of Chinese refugees from Burma, the school expanded quickly and by 1966 included a middle school section. However, since the school name "*zhongxing*" in Chinese implies the revival of Chinese dynastic domination, villagers feared Thai authorities might interpret it as having a pro-China political significance. To avoid this misunderstanding, villagers changed the school name to Yixin (or new) Middle and Elementary School (一新中小學校) in 1970.

This school began to receive formal assistance from the Nationalist government in Taiwan after 1982, a year after the Thai state granted citizenship to Maehong's military personnel and their dependents.The textbooks provided by the Nationalist government, with their strong anti-communist rhetoric and explicit respect for traditional Han Chinese values, are considered ideal curriculum material for the pupils. Besides supplying textbooks and on-the-job training for schoolteachers, the Taiwan government also provided fifty scholarships each year to graduates from Yixin Middle School, so they could continue their education in Taiwan. Economic prosperity in Taiwan during 1980s and 1990s meant that there were plenty of jobs available for village youth who completed their college or professional school education in Taiwan. Over the years, according to Mr. Duan, who was not only the first graduate of the village Chinese school but also the first Maehong youth to receive a baccalaureate degree in Taiwan, more than one thousand village youth are living and working in Taiwan today.

A critical blow was delivered to the village Chinese school in 1986, when the Thai state adopted a drastic nationalist policy banning education programs in the ever-expanding Chinese immigrant communities throughout Thailand. The booming Yixin Middle and Elementary School was taken over by the Thai government and converted into a Thai public school. Fearful that the Thai government might adopt even more drastic and discriminatory policies against them, the Yunnan Chinese in Maehong decided to lie low, registered no formal complaints, and simply moved the instruction into private homes. Furthermore, to avoid raising the suspicions of local Thai authorities, the village governing body, the Self-Governing Committee (SGC, to be discussed more fully in the next chapter), reported to the Thai government that their dispersed evening classes were organized to instruct villagers on new agricultural technologies. To avoid being detected by the Thai authorities, since they continued teaching Chinese language and culture, SGC posted sentries around the village outskirts to alert the entire community whenever unidentified outsiders approached. In such situations, the instructors and the students quickly hid their regular textbooks and replaced them with objects or tools related to agriculture.

This dismal situation continued for almost three years in Maehong and other villages in northern Thailand until the Thai government began to relax its control—a development that coincided with the rising importance of Taiwanese direct investments to Thailand under its Southward Movement Policy (南進政策), mainland China's rapid economic growth, and the subsequently expanded trade with Thailand. Given the increased commerce between Thailand and Greater China, the demands for bilingual interpreters increased dramatically. Under the pressure from the domestic business community, the Thai government began to promote Chinese language instruction in major universities; it also granted permission to build or rebuild supplemental Chinese schools at the K–12 levels.

The year 1989 marked a major transition in Maehong's history. That year, Mr. Huang, a 40-year-old of Maehong's second-generation Chinese, was appointed the head of the village self-governing committee (SGC, *zizhihui,* 自治會) by General Li Wen-huan. Huang, a middle school graduate, was ambitious, visionary, and a skilled entrepreneur. Most importantly, he maintained a close relationship with General Li's children. To demonstrate his capability and ambition, he proposed rebuilding Yixin Middle and Elementary School to replace the one confiscated by the Thai government. With wholehearted support from the majority of villagers, the school was rebuilt in 1990 at a new site north of the village; it had not only new buildings and classroom facilities, but also expanded curriculum programs—from the previous three divisions of kindergarten, elementary school (first to sixth grades) and middle school (seventh to ninth grades) to a new high school program (tenth through twelfth grades)—making Yixin School one of only two Chinese schools in all of northern Thailand to grant high school diplomas.

The Chinese school in Maehong is set up to accommodate and supplement the Thai public education system in both its class scheduling and curriculum content. The Thai public school runs from 8:30 in the morning to 3:30 in the afternoon, Monday through Friday. After finishing the Thai public school day, students go home for a little rest and simple dinner, and then head to the Chinese school for evening classes that begin at 6:00 p.m. and last until 8:00 p.m. On Saturdays when the Thai school is off, the Chinese school holds daytime sessions from 8:00 in the morning to 12:00 noon. In other words, students attending the Chinese school spend an average additional fourteen hours in class each week compared to students attending only the Thai public schools.

In terms of curriculum, the Chinese school especially emphasizes Chinese cultural heritage and practical Chinese language instruction, including reading, writing, and conversation. Culturally focused language instruction is considered the key subject and is taught from kindergarten to grade 12. Textbooks used in Chinese language classes are shipped directly from Taiwan, and they deliver heavy doses of traditional values like filial piety, honesty, and loyalty to the Chinese nation-state. The school has willingly paid higher salaries to attract top instructors of this subject.

Besides Chinese, the other subjects are mathematics and English, both considered important for graduates whether they go on to college in Taiwan or find jobs in Thailand. Middle and high school students (seventh through twelfth grade) also take computer classes—another pragmatic subject crucial to future employment.

Students attending the Chinese school pay a monthly tuition that increases from the lower grades to the higher ones. Kindergarten costs 500 baht per month (US$12.50 at the 2005 exchange rate), followed by first and second grades, which cost 600 baht (US$15). The third and fourth grades cost 700 baht (US$17.50), and the rate jumps to 1,100 baht per month for fifth and sixth grade. The monthly rate for middle school is 1,600 baht (US$40), and 2,000 baht (US$50) for high school.

Tuition differentials are reflected in teachers' salaries. A monthly salary of 3,000 baht (US$75) is the standard payment for teachers from kindergarten all the way up to fourth grade. Teachers in fifth and sixth grades receive 3,500 baht (US$87.50) per month. Middle school teachers' salaries jump to 5,000 baht (US$125) per month, while high school teachers command 6,000 baht (US$150). Those teachers having administrative responsibilities collect an additional monthly stipend of 1,000 baht as a form of compensation, and the highest is the schools principal's 8,000 baht.

Because most Maehong Chinese were military personnel or caravan traders when they settled in northern Thailand, few of them have the qualifications to teach in the Chinese school, especially at the middle and high school levels. One important source of teachers has been more recent refugee flows from the Mainland, both during the chaos of the Cultural Revolution (1966–76) and the more relaxed post-reform era since the 1990s when greater geographic mobility became possible in China.

Young Chinese, mostly men, with a high school or college education, have been slipping through the porous Sino-Burmese borders en route to northern Thailand where they continue to find Chinese schools eager to hire them as teachers. Some recent immigrants from China claiming to be college graduates have taught subjects such as geometry or advanced algebra, English, and Chinese at the high school level. Since few refugees arrive with diplomas or official credentials, their qualifications can hardly be verified. Nevertheless, their lack of credentials has posed no long-term problems for Maehong's Chinese school. And most refugees from the Mainland in fact have used teaching in Chinese schools like Maehong's to meet the short-term needs of shelter and income on the way to acquiring official Thai residence and moving on to big cities such as Chiang Mai and Bangkok (more in chapter 5).

If the teacher's monthly salary is not high, teaching is not exactly a full-time job either, since teachers typically spend only two hours in class each day during the week. Most teachers who are long-term residents in Maehong hold down other jobs to supplement their teaching salary. One married couple, both college graduates from Taiwan, teach high school Chinese and computer; they also own hill fields and a small computer training outfit. Another couple from Hong Kong, who teach high school English and computer, are actually Christian missionaries. A middle-aged teacher of elementary school Chinese is also the village postman and delivers mail six days a week. But most teachers who recently migrated from Mainland China do not have second jobs, either because they don't have the means to establish another line of business or because they don't consider residence in Maehong a permanent choice.

Over 1,400 students were attending Yixin School in the summer of 2007 when I last visited. Most are the children of village residents, but there are also a few from nearby Yunnan Chinese villages where there is no high school. I was told that half a dozen Thai children from nearby villages also attend. Thai parents send their children to Chinese school because becoming bilingual in Thai and Chinese greatly improves their employment opportunities in Bangkok and southern Thailand.

Many village parents use the employment conditions in Thailand to underscore the importance of Chinese education. A monolingual (i.e., Thai-speaking only) college graduate in Bangkok can at most command 4,000 baht (US$100 at the 2005 exchange rate) per month as a clerk at an import-export firm, while a bilingual college graduate can double that amount easily. Besides employment in trading companies, the booming tourist trade in southern Thailand has also created great demand for bilingual tour guides and office workers who can accommodate tourists from Hong Kong, Taiwan, and, increasingly in recent years, China.

The majority of the students at the Yixin Middle School attend classes willingly—they acknowledge the financial benefits a Chinese education may bring to them in the future and the sacrifices their parents are making now to cover their tuition and other expenses. There is, however, a small minority who resent the additional fourteen hours of schoolwork each week, especially among rebellious teens. Compared to the easy-going Thai teenagers, Chinese youth seem to be over burdened with classwork and more subject to restrictive parental control. Premarital dating and sex among teenagers is generally frowned upon by village parents, but village youngsters see plenty of sexual intimacy among Thai youth and on television. Several high school teachers complained to me about disciplinary problems in the classroom. When they impose discipline for open necking in the classroom, or punish female students for sitting on male students' laps during class time, the students sometimes retaliate. One student who physically attacked a teacher in 2004 was immediately expelled. To vent his anger, this student then planted two homemade bombs in his classroom in an effort to destroy the school. Other teachers complain of finding used condoms in the back of the classroom after the recess between classes. In general, however, such disciplinary problems have been the exception rather than the rule in the Chinese school.

REPRODUCING THE PATRIARCHAL FAMILY

The Yunnan Chinese in Maehong follow traditional Han Chinese kinship practices, which are predominantly partrilocal and patrilineal. Besides reckoning kinship ties, the patrilineal principle also dictates the passage of family name and property from father to son(s). Recorded family history in the form of genealogies called *zupu* (族譜) inscribe the male line in successive order; this systemically places women in subordinate positions that define them as a mere appendage of the patrilines, namely, daughters of their fathers, wives of their husbands, and mothers of their sons.

Anthropologists in the China field have reported the supremacy of the family (*jia*) ideology in people's daily lives (Cohen 1976). In the name of maintaining family integrity and continuity, women not only accept their subordinate positions in the gender hierarchy, they are also stalwarts of its reproduction. To fully explain how familism and, implicitly, male domination have been produced and reproduced in Maehong, I will now turn to the following three subjects: the construction of the family genealogy, the subordination of women in the domestic gender hierarchy, and the construction of the joint family.

Constructing the Family Genealogy

It is a common practice among Han Chinese, both domestic or overseas, to compile genealogy books once they have firmly settled in a locality and intend to establish a patrilineal kin group, called *zu* (族) or *zongzu* (宗族). A genealogy book can be described as a formal written record of the male lineal history, beginning either with a mythical founding father in ancient China or the founder of the current residential group, through all male issues in subsequent generations. The only females included are wives of the ancestors—since they beget male descendents. In general, the ancestors' sisters are not included in the genealogy, since they belong to the genealogy books of their husbands, although there are exceptions, as will be analyzed below (see Liu 1992: 5).

The genealogy book is considered one of the most important family heirlooms; it is generally passed from father to oldest son to ensure the continuity of the family's "incense and fire." The existence of the family genealogy provides a historical context within which the individual members anchor their temporal existence by its symbolic embodiment in the written record.

Not all Chinese families or households have their own genealogy books, however. Several preconditions had to be met before its compilation could be commenced. First, the family head (normally the father of the family) must have access to the genealogy from his natal home so a copy could be made as the basis for future expansion. Since most Maehong villagers were forced out of China by the Communist forces and, before the 1990s, they had only limited access to their natal homes in Yunnan, this prerequisite was not easily accomplished. Only after China entered its full-scale capitalist modernity period under the Reform and Open Policy of the 1990s did increased traffic between Yunnan and northern Thailand permit more and more of the village families to copy their genealogy books back home and compile new ones in Maehong.

Second, amendment of a family genealogy requires special writing skills—it should be done by a hired compiler skilled at literary Chinese who is willing to undertake it for a predetermined payment. Printing the new edition of the genealogy requires additional funding. Only families with substantial incomes and high social status can afford this endeavor. For most of the poor foot soldiers, barely able to sustain their own lives, the idea of compiling a family genealogy is beyond their dreams.

Symbolically, then, the family genealogy contains three important meanings. The first is the family's established status in the community: a genealogy indicates that a family intends to settle permanently in the village and thus establishes it as the founding nexus for future generations of its patriline. The second is the demonstration of a family's wealth: the lineage can afford to have its existence formally inscribed in written form for future generations to commemorate. The genealogy book that embodies this patrilineage in this very locality essentially collapses time and space into a single written representation (Casey 1996; Giddens 1991; Hall 1996). Lastly, the production of the family genealogy manifests the supremacy of the male gender over the female. Formal kinship links are demonstrated only through the males in both ascending and descending generations. Female existence, following the idealized Confucian familial practice, is tangential to a husband's temporal and spatial inscription in the genealogy. Without her husband, a woman has limited existence temporally, socially, or spiritually. As such, she will become a wandering ghost to be feared, pitied, and mollified in popular folk rituals (Li 1985, 1986; Weller 1987).

I use two genealogy books that I photocopied in Maehong in 2003 as examples in my analysis. Volume A is a hand-copied genealogy of twenty-three pages compiled by the household head, Mr. Peng in 1997. Volume B is a printed genealogy of nineteen pages, compiled by Mr. Li in 1996. Needless to say, both families are considered to be financially secure by village standards.

Volume A follows the standard format that begins with the legendary period of Chinese history (circa BCE 2205–2852), tracing the Peng ancestry to one of its heroes; the next thirteen pages record male issue from this mythical ancestor and his subsequent descendents in generational order; each entry mentions meritorious deeds performed, including special honors or official titles received, and the specific localities where these men established their households. This family originated in Jiangsu Province and is the direct heir of the legendary Peng Zu (彭祖), a Daoist immortal, who claimed to have lived for 800 years. One of his offspring branches moved to Jiangxi (江西) during the Tang dynasty (circa CE 618–905), and, several generations later, another branch moved from there to Yunnan during the Song (circa CE 960–1278). It was not until Mr. Peng's father's time that this family moved to Burma. Mr. Peng was born in Burma in1936 and he subsequently moved to Maehong in 1973. The genealogy book contains information about thirty-three generations: thirty-one generations before Mr. Peng and two after him. Mr. Peng also provides detailed information about his three wives, three sons, and four daughters. He explains the circumstances under which he married these three women, and why he disowned his second son and expelled him from the family because of his drug addiction problems.

Like most genealogy books in China, this one has a lot of omissions and fabrications. For the covered time span of over 4,000 years, each of the thirty-three generations must have lasted more than one hundred years, even if we take into account the extreme longevity of the original progenitor, Peng Zu. Historical authenticity aside, it does assert the status of the Peng family in Maehong and establishes Mr. Peng, the current household head, as the founding father of the Pengs.

The second genealogy is more modest in its claimed duration: it traces the Li lineage back to the Qing dynasty (circa CE 1644–1911) when it moved from Jiangxi (江西) to Yunnan. The compiler, Mr. Li, was born in Zhenkang County (鎮康), Yunnan, and his entire family moved to Thailand in 1949. This genealogy contains twelve generations, with Mr. Li in the ninth generation. One unique feature of this genealogy is its inclusion of some females and their spouses (but not their offspring since they don't have the surname Li) from the sixth generation downward, about the time the Li lineage moved out of China in 1949. This practice may be attributed to the following three reasons. First, this lineage group, being relatively small, had few male members, so the inclusion of married-out daughters and their husbands would increase its size and social heft. The common practice of uxorilocal marriage in Yunnan may also contribute to the inclusion of daughters and their married-in husbands. The second possible reason arises from the dispersion of lineage members over a wide geographic region—Yunnan, Thailand, and Taiwan. Recording married-out daughters and their husbands is practical because it provides a way to track them down in time of need. The last possibility, I speculate, is the demands of the changing times. At a time of "high modernity," characterized by the discontinuation of tradition and fluidity of individual identity (Casey 1996; Giddens 1991; Hall 1996), the patriarchal emphasis may be weakening as daughters' importance has increased in "post-traditional" society, as seen in other parts of China (Ding 2006; Tam 2006).

This "post-traditional" argument should not be taken too far, however, because this genealogy explicitly inscribes Confucian values in its "Family Teaching" section, with the opening exhortation of: *zhong* (忠, loyalty), *xiao* (孝, filial piety), *jie* (節, proper manners), and *yi* (義, righteousness). The notes below this "teaching" include three "musts" to which all descendents of the Li lineage must follow: They must be good sons and daughters at home, must be good students at school, and must be good citizens in society.

For these Maehong families, genealogy books provide the male patriline with historical authenticity and legitimacy. Males are anchored in time and place as their names are inscribed and manifest ongoing lineal connectedness. Second, their manner of inscription affirms the supremacy of the males over females; still they are not absolutely exclusive but flexible, adjusting to historical or other contingencies. And lastly, the family genealogy is a transmitter of ethical standards that are clearly Confucian in nature. This notion accords with de Bary's view that as China's modernizing project proceeded in the 20th century, the family would become the last remaining bastion of Confucianism (de Bary 1988).

The Subordination of Women in the Domestic Gender Hierarchy

When the Yunnan refugees established Maehong village in 1964, they were sustained by the ideological canon of traditional Han Chinese culture that had provided them a secure mooring during their protracted war against the Communists who occupied their homeland. As explained above, this reified Confucian ethos is fundamentally male-centered and so marginalizes women. This cultural ideal was further reinforced by the fact that the occupations of the early Maehong settlers were either military personnel or caravan traders—professions pursued exclusively by males. Even after the village was fully settled in the 1980s, the strong emphasis on military hierarchy preserved the gender imparity and essentially naturalized it. Members of the village Self-Governing Committee (SGC, 自治會) were always appointed by General Li, and to this day no woman has ever been appointed director or deputy director of that organization. At public events, such as village meetings, women were not expected to speak. Any woman who did would be shut down with the criticism: "Ignorant women folk!" (不懂事的婦道人家.)

In daily life, women's inferior social status is even more explicit. Incidences of domestic violence can only be whispered among the wives since they have no official authority to appeal to; such events are totally ignored by men as if they are the natural and normal state of affairs. Since a woman is a mere attachment of her husband in the patrilineage, she may lose everything, including the house she lives in and the farmland she works on, if her husband dies before she produces a son. The dead husband's brothers may arrive unannounced and seize all the family belongings from her since she is no longer a part of the patrilineage.

Similarly, in terms of behavior, there is an explicit double standard that favors men over women. Men can engage in illicit activities, such as drug use, gambling, drunkenness, visiting prostitutes, taking concubines, and even wasting family assets with impunity and without provoking public censure. But for women, any hint of misconduct or impropriety exposes them to the rage of public opinion and sometimes even punishment at the hands of the SGC in the form of public humiliation or penalty. As anthropologist Chang clearly points out, for women, "The meaning of marriage was based on obligations instead of love. These obligations required fidelity to their husbands irrespective of their feelings, filial obedience to their parents-in-law, and production of male offspring. These were obligations conforming to the teachings of propriety (*lijiao*). Failure to fulfill any of these obligations could result in punishment or abandonment by the family" (Chang 2005: 62).

Two specific incidents that occurred in the 1990s fully demonstrate this strict expectation and are still discussed, privately, by village women today. The first occurred in 1996 and concerned a village woman who was accused of being an adulteress. Her husband, a drug addict, provided no support to her and their five children from his military pension. To support herself, this woman used an old pick-up truck as a commercial vehicle to transport goods and people for fees. She also hired a male Yunnan Chinese driver from another village for the business. Later, it was reported to the SGC that this woman had an affair with the hired driver. The committee voted to take punitive actions and she was summoned to the SGC meeting where she was forcibly handcuffed and had her head shaved bald. She was then paraded around the village in a pickup truck with a loudspeaker denouncing her. In the meantime, her alleged lover was driven out of Maehong.

The second case also involves a woman who was accused of infidelity. In the view of most village women, this woman's case was even more outrageous than the previous one. This woman married her husband in the early 1990s and lived with her husband's parents. When her husband went to Taiwan to look for work he was jailed for six years for reasons that remain unclear. The woman stayed with her in-laws and took good care of them like a normal filial daughter-in-law should do. In 1997 her husband was released from jail and returned to Maehong. She was soon pregnant, but before the child was born, her father-in-law and husband died within a month of each other. She did her best to provide proper funerals for these two men, but in the meantime she had to fend off her husband's brothers, who tried to take away her family property. The birth of a son resolved that problem—she was the mother of a male heir who belonged to this patrilineage and hence was entitled to keep the family land and house.

Three years later, in 2000, when her mother-in-law passed away, she decided to remarry. Through a match-maker, she married a Burmese Chinese. This remarriage enraged the head of the SGC who accused her of being an adulteress. She was summoned to the committee meeting hall and handcuffed immediately. When charged with infidelity, she retorted: "I am a woman and need a husband to support me. Since my husband is dead, will any of you on the committee support me? If not, then why are you accusing me of being an adulteress?" The SGC head, further enraged by her outspokenness, slapped her face a few times. Then he turned around and asked other committee members to do the same. But the others seemed to have sympathy for her and did not join in. Since there was little support for the charge against this woman, the SGC head had to release her.

The absurdity of this gender-based domestic hierarchy is that husbands and sons are often indulged while women are given no room for error: they are expected to be filial to parents before marriage and to parents-in-law after marriage, remain chaste to their husbands, and be ready to make self-sacrifice whenever needed. The filial daughters or obedient wives whose debt-ridden fathers or husbands force them into assembly-line work or sell them into prostitution, as reported by early ethnographic reports in Taiwan and Hong Kong (Kung 1983; Salaff 1975; M. Wolf 1978), have their equivalents in northern Thailand too.

This male dominance, or the "patriarchal world" as Maria Tam (2006) calls it, is so overwhelming in Han Chinese culture that even radical revolutionary movements, including the Chinese communist utopianism that swept through the Mainland after 1949, have not budged it (Croll 1983; Johnson 1983; Stacey 1983; M. Wolf 1985). Nor has the otherwise seemingly invincible force of the capitalist market economy. Studies done in many parts of China indicate that even when women have gained greater social and financial autonomy in the prevailing industrial regime, they rarely question their inferior status *vis-à-vis* their male siblings or husbands (Chang 2005; Ding 2006; Kung 1983; Salaff 1975; Tam 2006).

The older generation women in Maehong are largely comparable to their sisters in modern China. Reduced socioeconomic circumstances, especially during the border crossing years of the1950s and 1960s, introduced much harshness into their lives and, correspondingly, thrust many additional responsibilities on them (Chang 2005). Even today, many women reminisce about how they single-handedly raised their young children and cared for aged in-laws by tending garden patches where they grew vegetables to feed the family, as well as raising fowl or pigs to supplement their husbands' meager military salaries. There are vivid accounts of their packing up all the family belongings in an hour or two when the order came to relocate to avoid encroaching enemy forces. These women have been enduring, resourceful, and tenacious in caring for their children and making ends meet while their husbands are away fighting or doing business—including engaging in the drug trade. Many of the village's founding mothers suffer additionally due to husbands, many of whom are drug addicts, physically violent, chronic gamblers, womanizers, and so on. And yet the women remain loyal to this gender-biased kinship ideology, faithful to their sometimes in-name-only marriages, and purposeful in helping to reproduce this same patriarchal world (Chang 2005; Tam 2006).

Constructing the Extended Family

In Maehong, following the reified Confucian ethical canon, villagers look upon the joint family as the most desirable form of living arrangement. The ideal joint family includes the founding father and the married-in mother, with all their male issues and their immediate families (again, married-in wives and children) residing in the same household, sharing the same family budget, partaking in family rituals, and remaining undivided as long as the parents are still alive. The ideal family, in other words, is organized by patrilineal kinship with more than two generations of agnates and more than one conjugal unit living together (Chuang 1985; Huang 1993). In this patrilineal domestic unit, all daughters will be married out to other families, and wives for sons will be brought in from other families.

Ethnographic reports in China indicated that this ideal often remains just that, rather than the reality of normal family living circumstances (Cohen 1976; Huang 1993; A. Wolf 1985). And as elsewhere, the inability to create an ideal joint family through the normal course of generational expansion is not due to lack of interest or effort by the parents. All parents, when they can afford it, try to keep their sons, grandsons, and all future male issue in the same household. But due to practical reasons, including sibling rivalry, conflicts between mothers-in-law and daughters-in-law or among the latter, or even the need to achieve economic diversification, the joint family eventually dissolves and fragments into multiple nuclear families—or a "stem" family plus several nuclear families. A stem family is the one in which the parents either reside with one of the sons permanently, or rotate among the sons' families following a predetermined schedule (Chuang 1985; Hsieh 1985; Huang 1993; A. Wolf 1985). The obligation to support the parents financially will be shared by all the sons who do not provide living quarters, meals, or other expenses to them directly, following a predetermined agreement often written in a contract.

In Maehong, the ideal joint family is actively promoted by parents whenever they are financially able to sustain it. However, many factors are obviously not conducive to the formation of a joint family. The lack of wealth to obtain or build a large enough house to accommodate the needs of an expanding family may force some of its branches to seek residence elsewhere. Dispersed residential arrangement is certainly not conducive to the cohesion and operation of the joint family ideal.

A related factor, commonly seen in Maehong, is that some sons may have moved to Bangkok or Taiwan on a long-term basis for schooling and later employment. They may eventually fully settle in these destinations, including purchasing their own houses or apartments, finding their own spouses, and managing their own family budgets. Interestingly, this type of dispersed family is considered undivided in the eyes of the community, and the parents may visit these branches and reside there temporarily. Most children, when employed, periodically send money home to their parents, especially if they have extra financial needs such as medical expenditures, tuition payments for younger children, or repairs to be made to the old family house. In reality, it is very unlikely that branch families in big cities in Thailand or in Taiwan will ever be incorporated into the original family again. *De facto* family division is thus accomplished through physically dispersing the children to various localities when they grow up. Similar development has been reported in Taiwan during its early industrialization period, called "federal family" (Chuang 1985) or "dispersed joint family" (Cohen 1976).

A common practice in Maehong, especially among relatively well-off families, is to summon one of the sons home, normally the oldest, to stay with and care for the parents when they become old. If this son is unmarried, the parents will use the pretext of finding him a wife and holding a wedding for him. Village parents still wield considerable authority in "arranging" wives for their sons. My estimate is about half of the marriages among the village youth are conducted in this manner. If the son is already married, the parents may pretend to be seriously ill and insist that the son and his family to come home to care for them. I met many middle-aged men in the village who complained about having been fooled by parents who played this trick. These men may have abandoned their careers or even loved ones in the city (either in Thailand or Taiwan) and returned home. Private grudges aside, the obedient son who fulfills the filial piety norm that Confucian ethics dictate is praised by the public.

In this type of stem family, the son who remains at home will inherit the family house, farmland and whatever property the parents may have in their possession. Once the parents have passed away, the other sons who live off in the city may return home to formally "divide the family." Generally, since life in the village does not offer good jobs that command good incomes, the returned sons do not actually ask for an equal share of the family wealth. The oldest (or youngest) son will inherit the entire family estate and become a caretaker of the ancestral home, property, and parental graves.

On major seasonal holidays, such as Chinese New Year (in late January or early February) or the Mid-Autumn Festival (*zhongqiu jie* 中秋節), in mid-September), or on the ritual occasions of Chinese popular religion, such as tomb sweeping on the fourth of April or the Ghost Festival in the middle of the seventh month of the lunar calendar (generally in mid- to late August), the urban branches of the family may return to Maehong and live in the village home for a few days. This type of family reunion may also occur on the parents' death anniversary when all the sons return home to pay their respects. These ritual occasions provide opportunities for the family to rekindle kinship ties. Since living in urban settings often means dwelling in high-rise buildings with only cramped space, these branch family members look forward to returning to the parental homes as an idyllic escape. Symbolically, they consider the parental home in Maehong the legitimate locus of both lineal succession and the dispersed family segments. It is here that both time and space collapse into the family "place," as Casey argues, "the phenomenological fact of the matter is that *space and time come together in place*. . . . Rather than being separate by equal cosmic parameters . . . space and time are themselves coordinated and co-specified in the common matrix provide by place" (Casey 1996: 36, italics original). The place, including the village in which they grew up, the parental home where they found shelter when young, and the graves where the departed parents are finally lodged, encompasses the lived experiences of the villagers who construct the meaning of life accordingly.

The irony of this traditional Han Chinese family system with its biased gender hierarchy is the active and key roles played by the subjugated wife (*vis-a-vis* her husband) and by the domineering mother (*vis-à-vis* her sons), never mind the formidable mother-in-law (*vis-à-vis* her daughters-in-law). Because they were literally uprooted from their original communities in China, Maehong's first-generation women never have the opportunity (or misfortune) of relying on their mothers' "uterine family" for the social or emotional support that Margery Wolf (1972) identified among rural Taiwanese women. And yet these women, despite their resourcefulness when crossing borders, rarely exhibit the characteristics that Chang described: "(when) confronted with the special circumstances, Yunnanese women were galvanized or pushed to transform their cognitive knowledge of traveling into corporal practices. The repeated movements have also empowered their independence" (Chang 2005: 64). I have yet to find a Maehong woman who is independent enough to openly defy the patriarchal family, the domestic gender hierarchy, or the patrilineal ideology that relegate her to an inferior position. On the contrary, when the subordinated wife or daughter-in-law achieves the status of domineering mother or mother-in-law, she becomes an active participant in and promoter of this male-dominant ethos, encouraging her son to return home by arranging a marriage for him, something that anthropologist Maria Tam also reported in Hong Kong (Tam 2006). The transition from victimized to victimizer must be triggered by major cognitive changes and self-justifications, a process that deserves careful research in the future.

POPULAR RITUALS

An important aspect of cultural reproduction resides in participation in popular rituals based on traditional Han Chinese belief systems—something akin to what we may call religious experiences. It is through this participation that the Yunnan Chinese diasporas refresh their collective memories of the cherished ethical canons in life, the epic heroes who exemplified Confucian virtues, and the reconfirmation of their ethnic identity as carriers of Chinese culture. Such rituals also provide the opportunity to relate their stock of acquired cultural knowledge to their offspring as they explain the multi-layered meanings of such practices. As Geertz points out, " . . . it is in ritual—i.e., consecrated behavior—that this conviction that religious conceptions are veridical and that religious directives are sound is somehow generated" (Geertz 1966: 28).

Popular rituals tend to be activities or performances faithfully reconstructed from the past, but with modifications or additions that reflect local or regional variability. When Maehong villagers reconstruct the spiritual domains of their ideal culture in this transplanted community, new rites and myths are added, and are inscribed onto the collective memories—a process that ensures every localized culture exhibits certain unique features that are not seen in other communities. I will discuss the formalized and not as formalized rituals in village daily life below.

Formal Religious Practices

The three institutions in Maehong that come closest to meeting the definition of "religious," as we ordinarily understand it, all have fixed locations in which ritual worship is held at fixed intervals. Marked by an architecturally distinct built environment that serves as the focal point of shared beliefs, they also command organizational support which maintains the spatial congregation and organizes activities for their followers. Furthermore, they represent three separate and distinct bundles of Han Chinese morals and spirituality.

Located on the western end of the village above the hot springs, the Guandi Temple (關帝廟) and Guanyin Monastery (觀音寺) are two grand buildings explicitly used for religious purposes; they were built by the villagers through public fund-raising in 1969 and 1981, respectively. These two religious establishments, both prominent in the villagers' day-to-day life, represent very different philosophical and spiritual concerns.

The Guandi Temple honors the ancient Chinese legendary hero, Guan Yu (關羽), a warrior general of the Three Kingdoms era (CE 221-263) skilled in martial arts and considered the embodiment of the Confucian morals of loyalty, bravery, and patriotism (Duara 1990; Feuchtwang 1992). In Chinese folk belief, Guan Yu is the patron deity of both military personnel and merchants (Feuchtwang 1992), and the cult of Guan Yu, wherein he serves as the village god, is widely observed across China (Duara 1990). Given Maehong villagers' strong military affiliation, it is understandable why this temple was built within a short five years of the founding of the village.

The major public festival at the Guandi temple is held on the twenty-fourth day of the sixth lunar month (approximately mid- to late July), Guan Yu's birthday. Additionally, villagers claim that this is also the day to celebrate Dimu (地母, Earth Mother, a local goddess in Yunnan) and Huoba Jie (火把節, the Torch Festival, an ethnic festival of several of the minority nationalities of western Yunnan such as the Yi, the Bai, and the Naxi). This compound celebration illustrates the syncretic nature of Chinese popular religions, which often merge officially sanctioned orthodox observances with those of local deities or cults that might otherwise be considered heretical by the court (Watson 1985; Szonyi 1997). In this case, the cult of Dimu and the Torch Festival, both carryovers from Yunnan's minority ethnic groups, were absorbed into the cult of Guan Yu in order to give them legitimacy.

Even though Guandi Temple is a formal structure that houses Guan Yu's statue, it has no resident priest or keeper to maintain the building. The tasks of temple management are carried out by officials in the village's Self-Governing Committee (SGC), which is responsible for collecting monetary donations for the temple from villagers and arranging Guan Yu's birthday celebration banquet at noon. On the morning of the birthday celebration, village women begin to trickle in at around 9:00 in the morning, bringing with them cooked food in baskets as sacrifice. They lay out the sacrificial foods in dishes before the statue, and then, each holding three lit incense sticks, they kneel in front of the statue, bowing or kowtowing toward it, then insert the incense in the incense pot, and go to burn paper spirit money in a brass urn out in front of the temple. After this ritual offering, they take the cooked food home to be used for that day's dinner, which celebrates the Torch Festival and the Earth Mother.

Meanwhile in the SGC compound, each village family sends a representative, usually the male family head, to take part in the public worship at noon. They donate 100 baht (or about US$2.50) as their contribution, and these payments are recorded by the SGC official for bookkeeping purposes. After paying their family dues, these mostly elderly men loitered around the SGC compound or walked over to the temple next door, greeting old friends and chitchatting about family events. The SGC hires half a dozen of the village women to prepare the luncheon at the temple. The two Guan Yu birthday celebrations I attended in 2003 and 2007 each had about 30 tables, representing over 300 participating families. Since the banquet is a lay religious congregation, the meal includes meat dishes and alcohol.

The Guanyin Monastery represents another interesting twist on Chinese folk beliefs. The deity honored here is Guanyin, also known as the Goddess of Mercy, patron deity of many Mahayana Buddhist monasteries in China, Tibet, Japan, and Korea. The common image of Guanyin in China is a female persona with numerous arms stretching from behind her back spread out like a peacock-feather screen. Guanyin is known as Avalokiteshvara, one of the bodhisattvas in the original Buddhist legend. In China Guanyin was worshipped as a male prior to the Song Dynasty (CE 960–1278). After that, the bodhisattva somehow acquired a female persona and new imagery, and became most closely associated with feminine compassion. Her willingness to shoulder worldly suffering underscores the morality of unselfish universal compassion. Guanyin is a popular deity in rural China, and many peasant communities take her as their local patron deity (Duara 1990; Feuchtwang 1992).

As a formal Buddhist monastery, the temple in Maehong has an ordained nun (*biqiuni,* 比丘尼) as its abbess with three female apprentices—all abandoned baby girls taken in by the nun. The nun is a Yunnan Chinese who was born and raised in northeastern Burma where she received formal training in a Mahayana Buddhist temple. Since Buddhism in both Burma and Thailand belongs to the Theravada variant, which is quite different from the Mahayana variant in China, there are few ordained Mahayana monks or nuns in the Golden Triangle. The Guanyin Monastery in Maehong recruited this nun only after considerable effort. Villagers told me that one of the early *phuyaiban* (the head of the village Self-Governing Committee), Mr. Duan, came from the same Burmese village where the nun grew up. Before the monastery was built in 1981, Mr. Duan visited the nun and promised that she would be the chief abbess if she decided to come.

Since the goddess Guanyin is the chief patron deity of this temple, worshippers observe the three major festivals that celebrate her life, in accordance with the lunar calendar: her birthday on the nineteenth of the second month, the day she achieved *nirvana* (*chengdao,* 成道), on the nineteenth of the sixth month, and the day she was ordained, on the nineteenth of the ninth month. In the morning of these days, almost all village families (except for those who have converted to Christianity) send representatives to the Guanyin Monastery to worship. On these occasions, villagers show their devotion by burning incense and kowtowing before her statue. They also make monetary contributions which are used to pay for the upkeep of the monastery and the noon banquet. Since Guanyin is a Buddhist deity, no meat dishes are ever served. Donations are again collected by an SGC official and turned over to the abbess, who also hires a few village women to prepare the vegetarian banquet for worshippers at noon.

The examples of the Guandi and Guanyin temples indicate that popular rituals can reinforce established cultural ideals among religious followers in the sense that shared experience in community events that idolize specific virtues upholds prescribed morals. This Durkheimian view, which may appear tautological at the first glance, actually insists that when people perform certain religious practices, they are complimenting themselves for the positive virtues that are crystallized in their religious symbols. These cultural ideals or morals, however, are not necessarily always something handed down from the past or imported from without. Local legends, either real or creatively produced, can sometimes be embraced and eventually incorporated into formal public domains. In so doing, local heroes or virtuous deeds become accepted norms, and will ultimately be standardized as part of the established official orthodoxy (Watson 1985; Szonyi 1997). It is with this understanding that we turn to the third formal religious establishment in Maehong.

Right next to the Guandi Temple on the western edge of the village is a small, inconspicuous shrine called the Local Master's Temple (*tuzhu miao* 土主廟), which was erected by villagers for worship of the village's unofficial patron deity. Even though this shrine is quite modest in scale when compared with either the Guandi Temple or Guanyin Monastery, it enjoys high popularity among villagers who flock to it on the morning of the lunar Chinese New Year's Day and on the eighth day of the second lunar month, the deity's birthday, to make sacrificial offerings. Inside this shrine is a statue made from earth or clay with black face and pre-modern-style military regalia. Outside the shrine stand two earthen statues in pre-modern civilian attire. My inquiry into the identity of the deity produced two slightly different accounts of a virtuous deed performed by a Yunnan Chinese soldier who was one of the earliest settlers in Maehong.[1]

In the first version of the story, when the Nationalist militias in the Baoshan region (保山) were driven out of China by Communist forces, a soldier surnamed Zhang left in such haste that his widowed mother was left behind. Once the Nationalists had settled in the Golden Triangle, Mr. Zhang, being greatly concerned about his mother's well-being, decided to sneak back to Baoshan fully armed, to fetch her. Since the only way to move across the hilly regions was on foot and because his mother was old and feeble, he had no choice but to carry her on his back on the way out. That was in addition to the weapon and food he had to bring for the journey. Even more dramatic, on the way out, their movement was detected by the Communist border patrol. Mr. Zhang was able to escape hot pursuit thanks to his good knowledge of the terrain and his bravery. Once out of China, he took good care of his mother until she passed away. To commemorate his exemplary filial piety, Maehong's villagers constructed this shrine for Zhang when he died and consider him the patron deity of the village, comparable to the community earth god in other parts of China.

The second version of the story also tells of Mr. Zhang's heroic and filial efforts to bring his mother out of China and is essentially identical to the first save for minor but telling details towards the end. As Mr. Zhang and his mother escaped from Yunnan they were protected by a local Baoshan patron deity, Tuzhu (土主). So it was to commemorate the good deeds of Tuzhu that villagers who came from the Baoshan region constructed the shrine after Mr. Zhang's death. No matter which version of the legend we follow, this process neatly summarizes Clifford Geertz's view: "In a ritual, the world as lived and the world as imagined, fused under the agency of a single set of symbolic forms, turns out to be the same world, producing thus that idiosyncratic transformation in one's sense of reality" (1966, p. 28). The mixture of human imagination with the 'real' in this legend, namely the consecration of Mr. Zhang's good deeds in the local myth, includes the delicate process of symbolically transforming the lived to the imagined and back, fusing ideals of Confucian ethics onto daily practices, and honoring a local patron god who serves as an agent that reproduces and reinforces those cultural ideals.

Seasonal Rituals

The second type of popular ritual in Maehong involves the seasonal changes of the calendar year. The Chinese way of calculating seasonality is complicated in that it involves the revolutionary movements of both the sun and the moon. The solar calendar, which divides the yearly circle of 365 days into 24 phases, or *jieqi* (節氣), is used primarily in weather forecasting and planning agricultural activities, as well as pinpointing the days of the spring and fall equinox (春分、秋分) and the summer and winter solstice (夏至、冬至). The lunar calendar, which follows the cycles of the moon, is used primarily for personal calculations, such as recording the birth and death day of an individual, and identifying the most auspicious time to hold specific events such as marriage, house moving, beginning a new venture, taking a trip, and so forth.[2]

The seasonal rites in Maehong involve ritual performances that mark the transition from one season to the next and accommodate the adjustment of organic humans, so that the vicissitudes of the universe properly correspond to the individual's microcosmic constitution. Using mostly the lunar calendar, we can identify the following major seasonal rites observed by Maehong villagers in chronological order. On Chinese New Year eve, which normally falls between mid-January to mid-February, all family members gather for the year-end meal. All those who work away from home will manage to return for the reunion. After dinner, family members pay their respects to the ancestral tablets first, and then to the most senior members of the family, such as the grandparents or parents. In more traditional families, children formally kowtow to their parents and grandparents, and the latter reciprocate by giving the former money, called *yasuiqian* (literally, 'suppressing the year' money) wrapped in red envelopes. After that, they light firecrackers to send off the old year and the family kitchen god, who is suppose to return to heaven and to make an oral report to the Jade Emperor about each individual family member's conduct in the past year (A. Wolf 1974).

In Maehong, the more formal part of the New Year's rite comes on the New Year's Day around noontime. Most village families prepare simple dishes and take them to the outskirts of the village to welcome back the family kitchen god. They then go to the Guanyin Monastery, the Guandi Temple, and the Local Master's Temple where they lay out the dishes in front of the deities and light incense and burn spirit money in sacrifice.

The second popular ritual in the first lunar month is the Lantern Festival (*yuanxiao jie* 元宵節), which is held on the 15th when the moon is full. This is the day when villagers send off their departed ancestors back to the spiritual world. For this event, many families confine themselves to a vegetarian diet for the day and make lanterns for children to play with outdoors at night.

The early third lunar month witnesses the Qingming Festival (*qingming jie* 清明節), the day to commemorate the departed ancestors by visiting and sweeping their tombs. This day belongs to one of the 24 solar "nodes," hence follows the solar calendar and falls on April 4. Villagers generally prepare cooked foods as sacrificial offerings, to be laid in front of the tomb. Besides tomb sweeping and food offering, villagers also burn incense and spirit money for their ancestors. Even if a family does not have an ancestral tomb in the village cemetery, it will still prepare food to commemorate the ancestors at home.

The fifth day of the fifth lunar month is the Dragon Boat Festival (*duanwu jie* 端午節), a special celebration allegedly established to commemorate Qu Yuan (屈原), BCE 340-278), a poet and patriot of the ancient Chu kingdom who drew and quartered himself when witnessing the decline of the kingdom. Village families prepare cooked glutinous rice wrapped in bamboo leaves for this day. Another common practice is for each family to clean the house and dust realgar power in its corners to drive away insects. Aromatic artemisia leaves are also hung over house entrance to keep insects away. In southern China, this festival is often accompanied by dragon boat racing in a river or a pond among neighboring communities, but this is not done in Maehong due to the lack of any significant body of water in the area.

The seventh lunar month is important because it is the month of the ghosts. Villagers believe that all the wandering ghosts emerge from hell on the first day of the month to find living substitutes so that they may reenter the reincarnation cycle, a belief that originated in the Hindu-Buddhist tradition (Weller 1987). A week before ghost month, village women begin to wrap gold colored papers into gold nuggets. By the fifteenth of the month, or the ghost festival day, most families will have accumulated large quantities of the paper-made gold nuggets that they store in bamboo buckets or cloth sacks.

The fifteenth day of the seventh month is the ghost festival; a public ceremony will be held outside the main house entrance or in the courtyard. Each family sets up an offering table upon which are laid out various types of food—some vegetarian and some meat dishes. Family members worship their ancestors first and then the wandering spirits with kowtows and by burning incense and spirit money. After that, all the paper gold nuggets will be burned either on the ground or in a metal wash basin. They believe that offered sacrifices will appease the ancestors and the ghosts so that they will not endanger living humans. Since the entire seventh month is ominous, villagers generally do not plan major events such as weddings, house building or moving, starting new businesses, or pleasure travel during this month.

The next major seasonal festival, the Full Moon Festival, also known as Mid-Autumn Festival (*zhongqiu jie* 中秋節), is observed on the fifteenth of the eighth lunar month. Since the full moon represents a perfect round shape, it also implies the reunion of all family members. Most Maehong children who work outside the village will make their best efforts to return for the family dinner. After dinner, family members sit outdoors to enjoy the full moon while consuming moon cakes, a kind of pastry with a baked flour shell and sweet bean-paste inside. Again, the round moon and the moon cake symbolize family unity.

The last major seasonal festival, which also follows the solar calendar instead of the lunar, is the Winter Solstice (*dongzhi,* 冬至), on December 21 or 22. The arrival of the winter requires nutritional supplements for all living humans, so that they may build up the physiological essence necessary to resist the expanding external cold elements. Besides meat dumplings (*tangyuan,* 湯圓), made of a rice powder shell filled with pork and vegetables and boiled in soup, families also prepare oily dishes such as sesame chicken or pork knuckles with beans for all family members; this is another occasion that reinforces family solidarity.

In addition to these traditional Chinese seasonal rites, villagers, especially the younger generations, also observe the Thai water festival, Songkran, in the middle of April. On that day, in anticipate of the arrival of the monsoon rains, people splash water on each other. While young people enjoyed the teasing and horse-play of tossing water, especially across the gender line, older villagers frown upon such activities. They argue that this is not a Chinese festival, and it might accidentally result in physical harm, as is reported in Thailand's newspapers every year.[3]

Seasonal rites are important among the Yunnan Chinese for two reasons. The first one is the explicit cultural ideals and moral values they carry. By partaking in these ritual activities, community members affirm their beliefs and commitments to those traditional ideals and morals. Certain traditional Chinese cultural beliefs, such as the sanctity of the family, the importance of ancestral veneration, the concept of the organic self versus the cosmic forces of the universe, or this world versus the supernatural world, are all embedded in these rites. Thus, when villagers perform seasonal rituals, they not only rehearse traditional cultural ideals but also reinforce their self-image as the embodiment of such ideals.

The second meaning of these rituals can be seen in their role as symbolic markers the Yunnan Chinese use for ethnic identity. Partaking in these popular rituals reaffirms their shared social experience and their commitment to the construction of an authentic Chinese community. The opposite of this type of affirmative social marker can be seen in the older villagers' reluctance to incorporate Songkran into their seasonal ritual routine. It is clearly worrisome to them that the younger villagers, in their readiness to be absorbed into mainstream Thai society, are willing to embrace this seasonal ritual as part of their lived experience.

Rites of Passage

The last type of popular rituals Maehong villagers regularly perform or participate in involves a person's transition from one stage of life to the next, such as school entrance, initiation into a specific organization, weddings, graduation, and funerals, all generally referred to as rites of passage in anthropology (Turner 1969; van Gennep 1960). This type of public ritual, using the words of van Gennep, marks "a passage from one situation to another or from one cosmic or social world to another," and can be roughly divided into three sequential stages: the rites of separation (or preliminal rites), transitional rites (liminal rites), and rites of incorporation (or post-liminal rites) (van Gennep 1960, pp. 10–11). Since these rituals cover a person's transition between two clearly defined statuses, its ambiguity and ambivalence (Turner calls it 'betwixt and between') also entails certain degrees of uncertainty or even danger. It is in this context that van Gennep used the term 'life crisis ceremonies' to address such events.

Different types and variations of personal rites mark an individual's changing life course in Maehong. For our purpose here, I will use two, the wedding and the funeral, to illustrate their significance, and for two reasons. The first is that these are the most public of all personal rites of passage, often involving large numbers of village families. As James Watson points out about late imperial China, " . . . rituals associated with marriage and death constituted a kind of 'cultural cement' that helped hold this vastly complex and diversified society together" (Watson 1988, p. 7). It is not uncommon to see dozens, if not hundreds, of gathered kin, friends, helpers, and guests who congregate for two or three days in the location where the main rite is held. Since a person's social position in Maehong, as in China, is judged by the number of people attending such events, and those who play host typically have their house gates wide open, it is easy for the visiting anthropologist to take part and to conduct interviews.

The second reason that I choose weddings and funerals as my analytic foci is because of the extent to which traditional cultural symbols are used, and ostensible changes, when detected, may signal people's shifting value orientations and personal identity. Weddings and funerals are the two most important life course events, and they have to be arranged properly and carefully. The selections of time and place, decorative motifs, costumes, music, colors used, procedures, and officiating personnel must be deliberated on carefully beforehand, and executed properly in due process to ensure the smooth transition of the involved parties.

Weddings

How a wedding is planned and executed is directly related to a person's status in life and his/her level of wealth. In Maehong, following the patrilineal principle, villagers generally hold wedding ceremonies in the bridegroom's house, but only for the first marriage. A widow or a widower who marries a second time will be most unlikely to have a large-scale formal wedding. Similarly, a poor farm hand without personal savings or a house to live in may not be able to have a fancy wedding ceremony. Two additional factors may also have a direct bearing on how weddings are planned. The first one is the sheer cost and scale of a wedding event. This may compel young people to rely on their parents' financial backing and management ability to arrange the wedding. Once parents seize the initiative, they may insist on following Yunnan Chinese wedding protocols with many traditional cultural symbols and practices attached.

The second factor that affects the villagers' wedding arrangements is their political allegiance to the Nationalist Party in Taiwan. The Nationalists launched a social movement called the New Life Movement (*xinshenghuo yundong,* 新生活運動) in China in 1930s, which aimed at eradicating many traditional customs and practices considered to be backward and uncivilized, such as bound feet (for women), opium addiction, arranged marriages, and taking concubines. In a wedding, the new morals promoted included freedom of choice for the couple and the wearing of modern Western attire (i.e., suits or tuxedos for men and white gowns for women). Another innovation introduced by the Nationalists was to have a senior person to officiate the process (*zhuhun ren* 主婚人), and two other respected persons, one serving as the official go-between (*jieshao ren* 介紹人) and the second as the witness (*zhenghun ren* 證婚人). Older Yunnan Chinese in Maehong have accepted this format and considered it the proper way to conduct a wedding.

Village parents consider many factors in selecting the spouses for their offspring; among the most important is ethnicity, and their preferences run as follows: Yunnan Chinese, Burmese Yunnan Chinese, Taiwanese, other Chinese Thai, ethnic Thai, and other hill tribes, in that order. Since I have elaborated on this topic in another context (Huang 2006), I will not dwell on it here. Other relevant factors in spousal selection are the two parties' relative age, education, income, professions, and family background. The younger generations tend to be more flexible toward opposite-sex friends who come from other ethnic groups, but pressures from their parents normally succeed in preventing them from letting such friendships develop into marital relationships. In the case of a marriage that involves a village boy and a woman from another ethnic group, there will generally be two wedding ceremonies, the first one following the wife's group's wedding protocol in her community, and the second a Yunnan Chinese one to be held in Maehong. In the case when a village woman marries a man in another village, following patrilineal principle, there will be no wedding ceremony in Maehong.

Wedding ceremonies in Maehong typically begin the night before the wedding day. The groom's family, when financial conditions permit, erects huge canvas tents in the family courtyard where many round tables and chairs are set out for visitors. The village Self-Governing Committee owns the tents, tables, and chairs that villagers can rent for such events. To one side of the courtyard the family also erects several temporary cooking stoves to prepare meals. The evening event begins with the eating of rice noodle soup at around eight o'clock. If the guests outnumber the available tables and chairs, they take turns. While some guests sit and consume their soup, others walk around, smoking cigarettes and chitchatting with the groom's family members or among themselves. The bride and groom also mix with the crowd to make their acquaintance.

After all the visitors have the opportunity to consume the late night snack, the wedding party begins its true festivities with a tradition called *dage* (打歌, literally 'hitting the songs'), a practice that seemed to have originated from some of the ethnic minorities in Yunnan. The middle ground under the tent is cleared by moving away all the tables and chairs. A group of middle-aged village men, about six or seven, each holding a three-stringed guitar, enters the open space playing their instruments, singing in unison, and walking in single file in a circle. The wedding songs typically instruct the marrying couple to live happily ever after by avoiding spousal discord through mutual respect, or to have sons as soon as possible to ensure the continuity of the family line, or to maintain proper personal conduct as filial sons- or daughters-in-law in their new life, and so forth.

As the musicians sing, they also match their dancing steps: three steps forward, two steps backward, and followed by a left kick and a right kick. When the singing and dancing begin, the future bride and groom immediately join the line, since they are the recipients of this public cajoling. Other bystanders first listen to the singing, and then gradually join in to follow the singing and dancing musicians and enlarge the circle.

While older villagers treat these traditional public rituals very seriously, the younger generations seems to take a rather casual, if not contemptuous, attitude toward them. At one of the weddings I attended in 2004, a young man in his twenties privately told me that both the musical instruments and the morals advocated in those songs are old fashioned and lack relevance in modern-day Thailand. He meekly suggested that if he had his own way, he would rather have modern karaoke songs or play western guitars than the traditional *dage*. However, he reluctantly admitted that since his parents would be most likely paying the expenses and organizing the wedding ceremony, he would probably have little choice but to go along with the dated rituals.

The wedding day begins at dawn for the newlywed couple, who under the guidance of the groom's senior relatives, make sacrificial offerings to the deity *wulaoshen* (五老神), the spirit match-maker for earthly humans worshipped in Western Yunnan. The couple will then pay respects to the groom's family by formally kowtowing first to its ancestral tablet, and then to the groom's parents and grandparents (if they are alive). After that, they greet other family members. Senior family members typically give gifts in the form of cash in red envelopes to the couple.

After the family ceremony, breakfast is served for friends and relatives who come from afar and stay overnight at the groom's house. By mid-morning, a reception table covered with red cloth will be set up at the entrance of the courtyard, and two family friends will be stationed there to receive well wishers and guests and record the gifts from them. These gifts are typically cash presented in red envelopes. The name of the guest and the amount he or she presents will be carefully recorded in the guestbook. This record will be kept for future reference when reciprocal gifts need to be sent, just as in many parts of China (Yan 1996). In general, the amount of cash sent to the couple reflects the closeness between the guest and the couple. The groom's maternal uncle may send 20,000 baht (about US $500) for this event. Ordinary villagers may come with cash gifts of between 200 to 400 baht (the first number must be an even number to indicate the pairing of the couple), ideally to pay for their personal cost at the banquet so the groom's family will not suffer any financial loss.

Typically, the luncheon or dinner on the wedding day is the major formal event. During the banquet, the couple and the groom's parents will come to each table to thank the guests. The newlywed couple will dress formally: the groom in dark suit or tuxedo, and the bride in white wedding gown. At some weddings close friends may serve as bridesmaids and best men for the couple. The three seniors who officiate at the wedding will take turns to make their public speech: these typically admonish the couple to be filial family members, to support each other until the end of their lives, and to have many sons to propagate the male family line. The same themes are prominently referenced in the colorful decorative emblems hung around the banquet rooms, with sayings such as '*bainian haohe*' (百年好合, a hundred years of harmony) or '*zaosheng guizi*' (早生貴子, have precious sons soon). In village wedding ceremonies, it is clear that certain traditional cultural ideals, such as filial piety, the importance of having a male heir to carry on the family line, and family unity, are still stressed and proclaimed publicly, even if with some reluctance by young generations.

Funerals

Just as when planning for a wedding, Maehong villagers determine proper funeral arrangements by considering relevant factors characteristic of the deceased or of the family head: his/her stage in life, wealth level, age, current and previous professions, and external social relations. In Maehong, older villagers generally insist on a couple of ground rules when arranging death rites. The first is, ideally, a person should die in his or her natal home. Thus, if an old man or woman is suffering from a terminal illness in a hospital in Chiang Mai or other city and is expected to die soon, his or her family will make every effort to prolong his/her life and charter a car to bring him/her back to the village. They believe that if a person dies in a distant land, his/her soul or spirit will not be able to find the way back and become a wandering ghost, a state dreaded by ordinary Chinese (Li 1985).

The second ground rule involves the timing of a funeral, which can be summarized as the 'three-day exemption rule.' When a person passes away, the funeral can be held at anytime within a three-day period without consulting the eight birth trigrams (*bazi* 八字) of the deceased or a *fengshui* master for the most auspicious location of the grave (*fengshui* 風水). If the funeral is held any later, however, the family must consult a geomancer to determine the most auspicious time (based on matching the eight birth trigrams of the deceased with the characteristics of natural time, based on lunar calendar) and the location for burial. To avoid these complications, most village families follow the 'three-day exemption rule.'

Besides these two ground rules, all other factors affecting a proper funeral are more or less personal. The cost of a funeral can range from several thousand baht up to as high as millions. Several key factors affect the costs. The first is the number of days of the funeral. The longer the formal mourning and funeral, the more expenses incurred since the mourning family must provide meals and lodging for visitors who come to pay condolences. Longer funerals also mean higher fees to priests, monks, and musicians who accompany the funeral rite. The second major cost factor in a funeral is the cost of the coffin and the tomb. A low quality plywood coffin, made locally, can be purchased for a few thousand baht. On the opposite end, a high quality teak or brass coffin may command up to a million baht. A respected community member, either the deceased or his/her surviving family head, may construct an extravagant tomb with above-ground cement cover and colorful ceramic tiles as decoration. Some well-to-do families even build a small pavilion near or above the tomb to mark its importance and to prevent weather-related damage.

The most costly funeral, villagers agree, is a traditional Chinese one. The public funeral rite may involve hiring Daoist priests to chant redemption verses and contracting musical bands for the funeral procession. Sometimes more than one geomancer may be employed to determine the most auspicious time for the funeral and the best location for the tomb. In addition, local craftsmen may be hired to construct miniature bamboo and paper sacrificial items, including a horse (to carry the deceased to heaven), houses, automobiles, and household amenities (to furnish the houses of the deceased in heaven). In a particularly lavish funeral I attended in 2005, in addition to the large quantity of paper money or paper-made gold nuggets and model household amenities, there were also model passports burnt for the deceased. Upon inquiry, the deceased woman's husband explained, "Since we live in this alien land, my wife's spirit may need passports in order to travel to Yunnan or Taiwan."

Some young village families, especially those with cross-ethnic marriages involving Thais, may invite Thai monks to participate in the funeral. There is no need to pay cash for their services, only gifts in plastic buckets, such as toothbrushes, toothpaste, soap, detergent, and so on, for personal use. The costs for Thai monks are thus lower than that of Daoist priests, possibly an incentive for the young generations to prefer them.

For funerals that follow the three-day exemption rule, the first day entails the announcement of the death, either by family members informing their close kin or by the village government through its broadcast system. The body will be dressed by family members in new clothing and laid immediately in the coffin that has been prepared in advance. Mourners must wear coarse cloth in different colors made from jute fiber. The coloration signifies the mourners' degree of closeness with the deceased (A. Wolf 1970). While family members prepare an altar in front of the coffin for the deceased (upon which is laid the temporary ancestral tablet, (*lingwei* 靈位), where the departed soul may rest, an incense pot, two candle holders, and other ritual paraphernalia, such as symbols representing other deities), some close kin, neighbors on good terms, or family friends begin to assemble to offer help with collecting necessary cooking utensils, borrowing canvas tents, tables, and chairs from the village government, and setting them up in the courtyard. Women help with folding the gold paper nuggets for the funeral or cooking in the kitchen while men run errands for last-minute items, such as candles, incense, liquor, and so on. If Daoist priests are brought in, they will be chanting around the coffin and guiding the mourners through specific rites.

The second day is for public condolences. Ordinary villagers bring incense and spirit money as gifts to the mourning family. After laying these items in front of the altar, the gift-givers pick three incense sticks out of the bag, light them with the candle, bow or kneel to the coffin, and stick the incense in the pot on the altar. As they offer their respects to the deceased, the mourners in funeral outfits, kneel around the coffin and bow back to the gift-giver in thanks. In the meantime, hired priests or musicians will be chanting or playing their instruments in the background.

The major mourning ritual is held on the third day so that the burial can take place in the afternoon before dusk. Visitors come with condolence money wrapped in white envelopes (as oppose to the red ones used for New Years and weddings), normally about 300 to 500 baht (or US $7.50 to US $12.50). In this case, the first number must be an odd number (i.e., 1, 3, 5, . . .), as opposed to the even numbers used in weddings. A reception table is set at the gate of the family courtyard where two villagers receive the money and record the gift-givers' names. Ideally, the gift money collected by the mourning family will cover the expenses of the funeral banquet. Just as with a wedding, the gift book is an important document to which the family makes reference when attending the funerals of gift-givers or their family members in the future. The rule of reciprocity is important, and the more people attending the funeral, the more face (*mianzi* 面子) to the bereaved family.

A luncheon is prepared for all participants. Non-vegetarian food is the norm, and with odd numbered dishes, in contrast to the even numbered dishes for weddings. After lunch, the funeral procession marches off, with a hired band of musicians leading the way. Some village families hire the village Thai school band that plays Western music, while others hire a traditional Chinese band that plays traditional music. For a well-to-do family, it may have both types of music. This is followed by the chanting priests, the coffin, which is carried by several hired workers, and the mourners in coarse jute outfits. Other visitors form the tail end, sometimes driving their automobiles to help the mourning family bring soft drinks, food, incense, spirit money, and so on to the cemetery.

For a traditional Chinese funeral, the grave yard is prepared by a few hired workers who dig the grave to accommodate the coffin. The work is supervised by a fengshui master (風水師), who instructs the workers when and where to dig following his calculations of the proper geomantic rhythms and the deceased person's eight trigrams. After the grave is dug, the fengshui master asks the offspring of the deceased to bury five auspicious items (including four plant seeds, rice, corn, peanuts, and dates, plus silver) at its four corners and in the center. All five items have the pronunciation of '*zi*' (子) at the end, and the five together make the auspicious Chinese expression '*wuzi dengke*' (五子登科), five sons who earned examination degrees from the court). After the burial, the offspring of the deceased connect all five spots with long cotton strings called connecting the dragon's veins (*jielongmai* 接龍脈), believing auspicious geomantic forces will actually be transferred to the offspring.

Before the coffin is lowered into the grave, the mourners lay out the food before it, including a bowl of rice, a few dishes (odd in number), a cup of tea, and a cup of wine. The mourners also light incense and stick it in front of the food and burn spirit money for the departed soul. The geomancer then uses a twig to symbolically sweep the coffin's surface before it is lowered. In a particularly dramatic funeral, I saw a daughter of the deceased woman, who was about thirty-some years old, jump into the pit, use her mouth to pick up a lump of earth, and spit it out above ground. Older villagers witnessing this act praised this young woman for her filial piety—she refused to accept the fact that her mother had actually departed, and was ready to use her own body to construct the grave for her mother.

After the coffin is lowered into the grave, the chief mourner (the oldest son of a deceased man or the husband of a deceased woman) hammers the last four spikes into the four corners of the coffin as the final seal.[4] After that, workers cover it with earth and begin to construct the round-shaped, above-ground tomb with bricks and cement. At the same time, the mourners take off the coarse jute cloth and set them on fire, along with the bamboo and paper horses, houses, spirit money, and paper household amenities they have prepared for the deceased. It is believed that the spiritual essence of the goods will go to heaven for the deceased to enjoy. After all the burning is done, the mourners and the others who accompany the family to the cemetery each eat a bowl of boiled egg in sugar soup (prepared by the mourning family in advance) to rid themselves of the polluted essence (*xieqi* 邪氣) associated with the recent death.

CONCLUSION

In this chapter I discuss how the first generation Yunnan Chinese diasporas in Maehong, as self-proclaimed carriers of authentic Chinese culture, attempt to construct or reconstruct traditional cultural ideals in the wilderness of northern

Thailand. As border crossing refugees, the Yunnan Chinese have already acquired a de-territorialized disposition through their earlier transnational movements. And yet their encounter with the Thai nation-state seems to have added to their sense of urgency about preserving their own cultural heritage and, furthermore, elevating their idealized version of 'Chineseness' to the status of an authentic ethnic identity. This paradox can be explained in this way: the modern concept of citizenship, whether based on a territorially defined political entity called the nation-state or a set of biologically endowed genetic diacriticals obtained through procreation, defines the traditional concept of 'Chineseness'—a cultural construction determined by the degree of conformity to Confucian moral ideals. In this context, Chineseness functions as symbolic capital with which the Yunnan Chinese anchor their identity to sustain and rationalize their marginal existence in Thailand's borderland. The acquisition of Thai citizenship provides them with full legal rights in their adopted homeland, yet it also creates a sense that their cultural heritage may fall away.

Through the construction of the Chinese school and the biased gendered hierarchy, and their participation in popular rituals, the older villagers declare their commitment to the cardinal morals and values of Confucian orthodoxy as they have understood them. Confucian orthodoxy, with its strong emphasis on a this-worldly orientation and graded interpersonal relationships, provides many of the guiding principles for the ongoing construction of the Han Chinese social universe. Reciprocal rights and obligations among family members can be best seen in the tenet of filial piety: the unconditional love and devotion of children to their parents and, reciprocally, the worldly security that parents provide their children. This theme is reiterated in almost all popular rituals, both seasonal and personal, as well as in the legend of the soldier Zhang who carried his widowed mother from China to Maehong in order to care for her. Filial piety also entails the importance of male issue to carry on the family line.

Closely associated with filial piety is the notion of family unity and harmony. In many seasonal rites and life crisis ceremonies, the importance of family continuity and harmony is reiterated again and again, as in folk advice given by the singers to newlywed couples in the *dage* performance, the public lectures delivered by respected seniors during wedding ceremony, and through the conspicuous display of auspicious symbols. Partaking in joint family meals not only confirms a person's membership in this "we" group, it also reaffirms his/her commitment to this moral ideal in the cultural reproduction process.

Ancestor worship provides an intimate spiritual world by which Han Chinese find consolation in the afterlife—an area that Confucianism is relatively mute about. However, as the Yunnan Chinese made the border crossing and settled far away from their old homes, their spiritual needs have changed with the new environment. The inclusion of passports for the deceased in ancestral rites so they can travel across international borders clearly indicates this new space-time dimensionality.

Popular rituals are important in that they provide many forums, public and private, in which the villagers formally profess their allegiance to this imagined cultural China (Tu 1994), affirm and reaffirm their adherence to its moral tenets (Huang 2003), seek spiritual tranquility in the afterlife through funeral and ancestral rites, and create new myths to glorify their otherwise mundane living.

The original Yunnan Chinese diasporas, through their rehearsal and participation in these popular rituals, seem to have been successful in constructing a Chinese identity based on shared social experience. Again, James Watson remarks on this issue: "To be a Chinese is to understand, and accept the view, that there is a correct way to perform rites associated with the life-cycle, the most important being weddings and funerals. By following accepted ritual routines ordinary citizens participated in the process of cultural unification"(Watson 1988 p. 3). To be able to correctly perform these rites, first generation Maehong villagers not only set themselves apart from the other ethnic groups surrounding them in this remote corner of northern Thailand, they also insist that they are members of an inclusive community who reproduce authentic Chinese culture. This argument may sound circular, but for the original diasporas their main concern is to satisfy the need of the inner self for an authentic identity and to achieve the goal of cultural reproduction to justify their marginal existence.

Despite such strenuous efforts committed by the original settlers, however, this cultural reproduction project appears to be fizzling out as internal conflict across the generations gradually intensified and many young villagers opted to move out and be assimilated into Thai society completely, or rebelled against the entrenched village authorities. This subject will be handled in greater details in the next chapter.

NOTES

1. Studies indicated that the worship of Tuzhu (土主) or Benzhu (本主) is actually a regional cult that commemorates the ancient creator of the Dali Kingdom in Erhai region, but was later converted into a local deity (Lien 2005).

2. The philosophical basis of the seasonal rites derives from what anthropologist Li Yih-yuan calls the human (i.e., Han Chinese) effort to maintain harmonious relationships in pre-modern China (Li 1993). He suggests that Chinese folk culture primarily concerns three levels of harmony: the heavenly (*tian* 天 i.e., the universe),

the human (*ren* 人 i.e., the physiological beings), and the social (*shehui* 社會, i.e., the human aggregates). Universal harmony is achieved through the delicate calibrating and positioning of human activities at the proper temporal and spatial junctures, as can be most clearly seen in the folk practices of fortune-telling and geomancy (*fengshui*, 風水). As for harmonious humans, a delicate balance must be maintained between the external, larger universe and the individual organic microcosm. The last level of the harmony, the social, involves the proper decorum among graded relationships that a person has in life.

3. I speculate that one possible grievance the older generation Chinese have against Songkran is the revealing of the female body when the water is splashed on the young women's and girls' light summer clothing. With its obvious sexual implications, it is considered immodest according to Confucian teaching about proper decorum across gender lines.

4. This practice is different from Watson's observation in that he suggests the final sealing of the coffin typically takes place before it is taken out of the community (Watson 1988 p.15).

Chapter 5

Emerging Internal Factionalism, Contested Political Loyalty, and Shifting Ethnicity

When the Yunnan Chinese first settled in Maehong, the Thai government applied its tribal policy in borderlands to this population as well; specifically, it appointed a *phuyaiban,* a village headman or chief to deal with public administration responsibilities. A great deal of political autonomy was given to the headman. This arrangement suited Maehong's military structure well, with its clear official hierarchy and unquestioned authority within the ranks. Under the watchful eye of General Li Wen-huan, the village Self-Governing Committee (SGC) was established; its membership was supposed to be elected freely by all villagers, but members were predictably all high-ranking officials, caravan traders or their offspring appointed by General Li. The committee then elected a *huizhang* (committee head or director), whose position was commensurate to the *phuyaiban* under the Thai administrative system for hill tribes.

The SGC was the de facto administrative organ in Maehong for almost four decades after its inception: it assigned tax responsibilities to individual families, collected taxes to pay for public expenditures, provided guidance for community development projects, kept the water supply system and garbage collection in operation, operated the village market, arranged festivals for major holidays, maintained the local Chinese school and a movie theater, paid the Chinese school teachers' salaries, and presided over internal disputes and dispensed justice as it saw fit. With its long military tradition, the SGC head ran the village like an army unit and reported only to General Li. Orders were issued from the top, and public policies were barely discussed before they became formal pronouncements.

Throughout the 1960s and 1970s, when the village had a more or less homogeneous population of military personnel and their dependents, the authority of SGC went unchallenged in Maehong. Under this system, the three most powerful persons were the head of SGC (the *phuyaiban* on the Thai government roster), who reported directly to General Li; the principal of the Chinese School, who handled the day-to-day functions of the school; and the chairman of the board of directors of the Chinese School, who also doubled as the SGC treasurer.

In 1989 when Mr. Huang was appointed the head of SGC, he appointed two of his close allies to form the ruling tripartite: Mr. Yang became the school principal and Mr. Liu the chairman of the school board. Mr. Huang, whose father had been division commander in the Third Field Army, retained close ties with General Li Wen-huan's family. He himself joined the military after middle school at the age of seventeen in 1966, and became friends with General Li's children. Huang's business acumen attracted General Li's attention and he was soon asked to run several investment ventures owned by General Li in the early 1970s, including the booming potato trade and mining in Burma. By the time the Thai government granted citizenship to the Yunnan Chinese soldiers in 1981, Huang, then in his early thirties, had established a food processing business with a couple of Taiwanese investors. He now owns a canning and food processing factory in Maehong and has become probably the richest man in the village through his successful business deals in and around the area. Since his food processing company hires over three hundred workers during peak season, he is also the largest employer in the village.

Huang, like any successful entrepreneur in East Asia, drives a Mercedes Benz and plays golf in Chiang Mai and other places with his business associates and military brass. His broad-based social networks allow him to represent the village in external affairs and take advantage of new developments or policy changes in Thai society. Initially hand-picked by General Li, Huang enjoyed almost ten years of supreme power in the village, until his fall-out with General Li's son in the late 1990s.

Huang relied on his two close allies to run the village. Mr. Yang, the school principal, had been a caravan trader heavily dependent on the military. After the disarmament of the Yunnan soldiers in 1981, Mr. Yang turned to farming and cattle raising, from which he has made a fortune. He owns a dairy farm with 110 head of cattle, and over four hundred *rai* of litchi orchards. He uses the grass he cuts from the orchards to feed the cattle, and the cattle dung to fertilize the fruit trees. He told me in 2004 that his best cows produce about 30 kg of milk each day, or over 10,000 kg per year. At the price of 10 baht per kilo, each top quality cow provides over 100,000 baht of gross income annually. Even though only 45 of the 110 cows are producing, the gross income each year can be as high as 4.5 million baht (about US$112,500). After deducting labor costs, feed, transportation, and medication, his net average yearly income is over 2.5 million baht, a respectable sum in Maehong.

Of Mr. Yang's five children (three sons and two daughters), four live in Taiwan; the oldest son is a physician and the second son an attorney. They are both married and intend to live permanently in Taiwan. The two younger daughters have also completed college education in Taiwan. One is a middle-school teacher, and the younger one is fresh out of college and looking for job. Only the youngest son and daughter-in-law remain in the village to tend to the family business. Mr. Yang's two sons in Taiwan donated a big sum of money for a new building in the Chinese school that contains half a dozen classrooms and the administration office. This donation, they declared, was an expression of their support of the school. As principal of the Chinese school, Mr. Yang is in charge of the cultural reproduction tasks that many Yunnan Chinese diasporas consider their most critical ideological commitment.

Mr. Liu, the chairman of the Chinese School board, was Mr. Huang's business rival in the early 1970s when potato production and trade were hot in northern Thailand and many Chinese diasporic communities were heavily involved. For a while Mr. Liu was the top potato wholesaler in northern Chiang Mai. However, after several years of expansion in potato planting, the inevitable potato blight and rust hit northern Thailand and caused widespread crop failures. The departure of U.S. military forces in Vietnam further contributed to the decline of the potato market. Mr. Liu pulled out of the trade completely and began to concentrate on building his orchards around the village.

As the chairman of the school board, he also serves as the SGC treasurer, handling all public funds. Mr. Liu also has five children, one son and four daughters. All his daughters are in Taiwan, either working after they finished college or, in the case of the last daughter, still completing her college education in 2007. His only son, in his early 30s, has remained in the village along with the daughter-in-law and two grandchildren. A successful businessman, Mr. Liu manages the public funds well and administers the daily activities in the village.

The tripartite rule of the SGC was smooth and successful for over ten years, from 1989 through 2000, when Maehong successfully rebuilt its Chinese school and made it one of the only two Chinese schools in northern Thailand that offer high school diplomas. But with the passing of General Li in 2000, and with the continuous influx of new immigrants to the village, the old military order has been gradually diluted. Furthermore, the internal diversification of the village in terms of wealth and ethnic composition has created friction and ever-widening resentments.

Another cause of internal conflict in Maehong can be traced to a policy change of the Thai nation-state. In an effort to incorporate the borderland regions into mainstream Thai society, the Thai government announced the establishment of direct election systems at the village level to elect village heads and representatives in 1997. Initially, to smooth the transition, the Thai government asked the *phuyaiban* to continue serving as village head until direct elections could be established in 2000. In Maehong the younger generation of villagers, with their Thai education and willingness to be incorporated into Thai mainstream society, began to challenge the old oligarchy associated with the village's military past. They pushed for a new community agenda aimed at re-orienting the village's emphasis toward Thai national priorities rather than preserving an imagined Chinese culture and identity.

In early 2002 the Thai government formally extended the civilian rules to tribal areas, including Maehong. This transition required each village to transfer its administrative power from the *phuyaiban* to a newly elected village representative committee, which, in turn, would elect a village head and three deputy heads to administer village affairs. From that point on, the Thai national treasury would tax citizens directly and provide direct subsidies to village governments for public administration and salaries for village officials. The village's former autonomy was stripped away and replaced with citizens' rights and obligations explicitly seated in the Thai nation-state and its constitution.

Maehong villagers in their thirties and forties who had received a formal Thai education and could read and write Thai welcomed such a development. They considered the old SGC oligarchy conservative and undemocratic. But this new development constituted a frontal assault on the authority of the older generation Yunnan Chinese diasporas. Still, since the change had been initiated by Thai government, the SGC could not openly defy it.

So a village-wide election was held in Maehong in mid-2002 and a village committee was formed, with one village head, a Mr. Zhou in his late thirties, and three deputy heads all charged with various aspects of public administration. With the successful extension of nation-state power into the village, the Thai government allocated a generous one million baht (about US$24,000) to the newly elected village government for public construction projects and building a new village government office in the middle of the village.

But when Maehong's newly elected village officials demanded the transfer of village funds and authority from the SGC to the village government, the head of SGC, Mr. Huang, refused. After prolonged negotiations, it was agreed that after January 3, 2003, the village government would take over most of the administration functions from SGC, such as maintaining the water supply system, garbage collection, public security, the weekly village market, as well as the hot spring facility and movie theater the SGC used to run. For a while Mr. Zhou, the village head, even demanded that village government take over the Chinese school, because it was under SGC control and was widely believed to have generated huge profits. Under tremendous pressure from villagers who had grown increasingly impatient with Mr. Huang's rule, all three power holders in the SGC were forced to resign and its operations were ceded to the village committee in May of 2003.

The new village committee head, Mr. Zhou, had a checkered personal history of drug trafficking and only minimal education, villagers claimed. After being elected as village head, he mistakenly believed that he had become the new *phuyaiban* with unchecked authority to pursue his own interests. He set up a private security force to patrol the village and intimidated those who dared to challenge his authority publicly. His abusive use of power soon brought him trouble. Late one evening in August of 2003, Mr. Zhou's night patrol arrested a village teenager on suspicion of using drugs. This youth was the son of Mr. Zhou's archenemy, so seizing the opportunity, Mr. Zhou had the lad severely beaten up. When the case was reported to the police in Chaiprakan County the next day, Mr. Zhou suddenly realized that he was not really the village *phuyaiban* after all, and that he had exceeded his legal authority under Thai civil law.

Even though bribery could easily make the charges go away in the Thai legal system, in this case the former SGC officials were ready to exercise their power and wealth to bring Mr. Zhou down. Mr. Zhou ultimately managed to escape criminal charges, but he was forced to publicly repent at a village meeting and return confiscated properties including the Chinese school and the movie theater to the SGC—the latter having been reorganized as a nongovernmental organization.

This incident highlights intergenerational conflict between the old Yunnan Chinese settlers and the younger generations, who are better at dealing with the Thai government. Looked at another way, it reveals the at least partial success of the Thai nation-state's policy to implement modern citizenship concepts in the tribal areas. The reduction of local or tribal autonomy and its replacement with Thai constitutional rights and duties point to Thailand's interests in incorporating all minorities into the modern nation-state (Keyes 1987; Wyatt 1982).

The building of a modern nation-state may have come late to Thailand, but it will fundamentally change the social relationships between mainstream Thais and minorities, as well as people's concept of citizenship and constitutional rights and obligations. The promotion of citizenship in Thailand with contractual constitutional rights and obligations will subject the Thai government, its bureaucracies, the police, and the military to closer scrutiny by common citizens, who will be more likely to assert their personal rights under the constitution and demand full rights as citizens.

CONTESTED POLITICAL LOYALTY AND CHANGING ETHNIC IDENTITY

Yunnan, in a remote corner of southwest China, has twenty-six ethnic groups. Ann Maxwell Hill's remark that Yunnan Chinese moved from a multi-ethnic environment in China to another multi-ethnic society in northern Thailand is insightful (Maxwell-Hill 1998). The term "Chinese," however, may be more complicated for the Yunnan Chinese in this context. Many Maehong villagers acknowledge that, during their grandparents' or great-grandparents' time, *tusi* (土司, the local tribal chief whose authority was acknowledged by the court in imperial China) still controlled their home areas; this seems to suggest that in this ethnically mixed area, some of their recent ancestors probably belong to Yunnan's minority nationalities. Even though today we often find in villagers' family genealogy descriptions of how and when their ancestors moved from inner China to Yunnan and subsequently migrated from China to Burma or Thailand, we also know that forging one's ancestry to claim legitimacy as a genuine Han Chinese is a common practice among sinicized or semi-sinicized minority ethnic groups, especially in southwestern China.

Once the Yunnan Chinese crossed the international border, old ethnic divisions seem to have subsided, and the term "Chinese" gained new meaning when confronted with new ethnic divisions or categories. They all became "Yunnanese" in the face of other Chinese immigrants, such as the Teochiu people. Both groups become "Chinese" *vis-à-vis* the Thais or other local hill tribes. An old Maehong villager whose surnamed He (和) signals that he is a Naxi minority from Lijiang, the center of the Naxi nationality, was not pleased when Professor Yang asked him if he was Naxi. He claimed that the Naxi nationality is a fabrication of Chinese Communist policy. He said that when he was young and traveled to various parts of China, he always referred to himself as a Yunnanese (*yunnan ren*) from the Lijiang area. It was only after the establishment of the Communist polity in 1949, he vehemently argued, that ethnic divisions became consciously constructed, and Naxi was reified as a new ethnic label for people in the area of his hometown.

Mr. He's position reflects traditional Chinese perceptions about ethnicity or race: that it is something tenuous and transitional, contingent on the degree to which individuals conform to Confucian ideals of civility (Harrell 1999). The advancement of the nation-state concept in the 20th century—with bounded geographic confines and clearly defined rights and obligations associated with citizenship—created a new sense of identity for these border-crossing refugees. It is in this context that the term "Chinese" obtains its new meanings for the Yunnan Chinese in the Thai nation-state.

Villagers claim that when they first came to Maehong in the 1960s, they encountered various hill peoples such as the Lisu, Lahu, and Akha, who practiced swidden agriculture in the vicinity of the village, and lowland paddy farmers such as the Karen and Thais. Once a permanent settlement was established in Maehong, most hill farmers sold their plots to the Chinese and left the region entirely. When I asked Mr. Liu, my key informant, why the hill tribes left this area when Yunnan Chinese settled here, he responded with a casual yet typically ethnocentric remark: "This is standard *hanlai yimie,*" (漢來夷滅), literally, the arrival of the Han Chinese signals the termination of the barbarians).

Even at the early stages of their settlement, the Yunnan Chinese migrants began to tap into their "Chineseness" by establishing ethnic networks beyond the village's vicinity and with other Yunnan diasporas communities, as seen in their expansion of potato farming and trade in the 1970s. Villagers reached out to Teochiu people in Fang, approximately 30 kilometers to the east; many villagers have Teochiu business partners and friends, and their children marry Teochiu people. Despite recognized differences—the Teochiu language and Yunnanese are mutually incomprehensible, and Teochiu people are generally better established economically—the Yunnan Chinese readily refer to the Teochiu people as "Chinese," just like themselves. This connection has worked in Maehong's favor, with the sharing of agricultural knowledge, information, and market connections described in chapter 3.

More recently, villagers cultivated close contacts with the Nationalist Chinese in Taiwan, especially since 1981. This also has had its advantages, with the Taiwan government channeling funding for public construction to the village, and village youth going to Taiwan for education, employment, and to settle with spouses and raise a family. A few Taiwanese have followed their spouses back to Maehong in recent years, adding to the complexity of the ethnically Chinese mosaic.

With its proximity to the Thai-Burmese border, Maehong has become home to many Burmese refugees. This includes not only Burmese hill tribes, such as the Shan, Wa, Lahu, Karen, and so on, but also large number of Burmese Chinese. The Burmese Chinese claim to be descendants of the last Ming emperor's followers who fled to Burma in the mid-17th century to escape the triumphant Manchu armies. These refugees formed many Han Chinese enclaves in northeast Burma and maintained Chinese language, education, and cultural traditions.

Given this ethnic mosaic, it was inevitable that the Yunnan Chinese diasporas would develop a kind of ethnic hierarchy, and nowhere is this hierarchy shown better than in male villagers' marriage preferences, both for themselves and for their offspring. In casual conversations, older village men often talk about who has a better marriage partner. Ideally, marriage partners are Yunnan Chinese who speak the same dialect and maintain similar customs. Most high-level officers in the Third Field Army and the rich caravan merchants who supply the military have educated Yunnan Chinese wives from good families. The second choice is Burmese Chinese, who also speak a Yunnan dialect and can read and write Chinese. Next comes the Teochiu people and other overseas Chinese, including Taiwanese. Even though they speak different "dialects" and follow different customs, they are nevertheless still "Chinese."

If all these options are exhausted, village men will consider marrying women from other groups in adjacent areas. In this case, Thais are preferred over "hill tribes." Thai women tend to be better educated and can provide good entry into the local society. The least desirable options are women from one of the hill tribes. Only villagers who have no other choice will consider marrying a Shan, Lisu, Akha, Wa, Lahu, or Karen. The negative view of hill tribe people was made clear in a comment that I heard while attending a banquet in Mae Salong. Two middle-aged men, both of whom had escaped from China during the Cultural Revolution (1966–76) and worked as schoolteachers in northern Thailand's Chinese schools, were lamenting the capriciousness of fate. As they parted, one consoled the other: "Hey, cheer up! At least we didn't resort to marrying Akha girls!"

This kind of negative stereotyping even runs across the generations in Maehong. In 2004, when conflict between the SGC old guards and the newly elected village governing committee deepened, I was puzzled by the bitterness of the old guards toward the younger village mayor, Mr. Zhou. They accused him of having made a fortune by smuggling heroin from Burma to Thailand. They also claimed he was once arrested for drug trafficking and spent four years in jail. According to Thai law, his criminal record should have deprived him of the right to be elected village mayor, an SGC official insisted. But Mr. Zhou managed to bribe court clerks to erase his criminal record, and with a clean slate he now pretends to be a decent businessman. Furthermore, they suggested, Mr. Zhou is a vicious person. He can talk sweet and appear to be sincere and honest, but then stab you in the back when your head is turned.

To me, these accusations made little sense, given most villagers' own checkered history of drug involvement and all sorts of backstabbing that had happened in living memory. My puzzlement was finally answered one day in August of 2005, when Mr. Yang rushed over to Mr. Liu's guestroom where we were visiting about another matter. Mr. Yang exclaimed, "He did it again, that bastard Zhou! Only someone of truly miserable character could do such a dirty thing! You know where his personal flaws come from?" Yang turned to me, "It's from that low-ranking foot soldier of a father and his illiterate Lisu mother!" Apparently the village's internal turmoil is more than intergenerational; it also involves class differentials and ethnicity.

CONTESTED POLITICAL LOYALTY

Since the first generation Yunnan Chinese in Maehong were driven from their homelands by Chinese Communist forces, they have remained strongly anti-communist and loyal to the Nationalist Party in Taiwan. With Nationalist government assistance to the village and many village children going to Taiwan to attend university and settle down, SGC officials even today bow to a portrait of Dr. Sun Yat-sen, the founding father of Nationalist China, and to Taiwan's national flag in important public ceremonies.

Recent changing political realities on Taiwan have tainted the Yunnan Chinese's views, however. The seizing of political power by the pro-independence Democratic Progressive Party (DPP) in Taiwan's presidential election of 2000 brought with it a drastic decline in financial support. The DPP government disbanded the office of the "Mainland China Relief Assistance Bureau" (大陸救災總會) that had been in charge of channeling funds to the village and other Chinese diasporas communities. By doing so, it severed formal official linkages between Taiwan and the Yunnan Chinese communities in Northern Thailand. Financial support to the Nationalist veterans in Maehong was transferred from the Taiwan government to a private religious charity organization, the Buddhist Compassionate Society (*Ciji gongdehui* 慈濟功德會), which provides a monthly meal allowance to the veterans, as well as medical assistance. Now, the only formal link between the Yunnan Chinese and Taiwan is through the college entrance exam. High school graduates from Maehong (and other Yunnan Chinese villages in northern Thailand) can take part in Taiwan's college entrance exam and receive some financial support if they pass. In winter of 2002, of the 37 high school graduates from the village, thirty-one passed the exam and went to Taiwan.

The reduced financial commitment from the Taiwanese government has angered many old diasporas in Maehong. They feel betrayed by Taiwan's government and by the Nationalist Party. They have openly expressed their contempt and criticism of the DPP, the Taiwanese independence movement, and Mr. Chen Shui-bian, Taiwan's former pro-independence president. One villager even stated, "If Taiwan ever declared independence, I would volunteer to lead a battalion of soldiers from China to invade Taiwan."

The changed relationship between the Yunnan Chinese diasporas and Taiwan has cast new light on China and contributed to changing attitudes toward communism. Older villagers are still avowedly anti-communist, but they are now more willing to examine the human/humane face of the Chinese Communist polity. The "reform and open door" policy of the past three decades, plus the astronomical economic growth in China, made them aware of and appreciative toward improved living conditions among their relatives and friends in Yunnan. Almost all of the older villagers have taken vacations to Yunnan in the past few years, and they talk about the positive impressions these made and the enormous changes in China.

The more entrepreneur-oriented villagers are excited about potential business opportunities with China, especially with China's new Mekong River Development Plan, which enhances river transportation between China and the nation-states along the Mekong River, including Burma, Laos, Thailand, Cambodia, and Vietnam. The provincial government in Yunnan, realizing the economic strength of Yunnan Chinese communities in northern Thailand, has been actively cultivating positive relationships with them by issuing special invitations. The thawing of animosity between the Chinese diasporas in Thailand and China will continue as long as China maintains its current political trajectory away from communism and toward an open and democratic society.

Despite China's improved image, most of Maehong villagers would not consider going back there to live for good. They have clearly stated that they appreciate King Bhumibol Adulyadej of Thailand for accepting them and granting them citizenship when they had become stateless refugees in the Golden Triangle. They also point out that, even though China is changing quickly, the Chinese diasporas in northern Thailand still have a better life than their relatives and friends in Yunnan. Even the winter weather is milder in Thailand than in Yunnan.

Village officials have taken every opportunity to express their gratitude toward King Bhumibol Adulyadej. The archway leading into Maehong carries the motto in carved Chinese characters: "The King's generosity and graciousness toward us can never be repaid." In public activities, officials always mention the King's kindness and vow to be good citizens of Thailand to repay the favor.

Rhetoric aside, the older villagers readily point to the corruption of local officials and the police. To keep the Border Patrol Police (BPP) from searching Maehong for illegal immigrants too often, the SGC used to provide monthly under-the-table payments of 10,000 baht (or $250) to them: 6,000 to the police station, 3,000 to the immigration office, and another 1,000 to the police overseeing village affairs. Although the Thai military is not as corrupt, this does not apply to the top brass. For instance in early 2002, the Thai army, under the pretext of eliminating opium traffic, seized the richest villager's property. Only great effort, including probably bribes delivered to top generals in Bangkok, got the seized property returned. Needless to say, in light of their precarious situation, few first generation migrants are willing to make a strong commitment to either Thai society or its citizenship. As transnationals, they are prepared to move on to a new locality or new nation-state with better living conditions whenever such opportunities present themselves.

ENGAGING THAI SOCIETY AND THE TRANSNATIONAL ARENA

The first generation Yunnan Chinese migrants generally have ambivalent feelings towards the Thais. On the one hand, with a certain degree of contempt, they consider Thais to be happy-go-lucky, hedonistic, and self-centered, and say they normally take a relaxed attitude towards work and life. When not pressed for cash, Thais prefer spending time enjoying life rather than working for more income or for future gains. Thai men drink when taking on laborious or difficult tasks, and by the time the drudgery of the work is done, they will be more or less drunk. Similarly, villagers say that Thai women crave material goods and comfort more than they care about familial well-being. If the husband can't meet his wife's demand for luxury items, she will leave him for someone more capable of doing so. They claimed this is definitely one of the major factors contributing to high divorce rates among Thai or Chinese-Thai couples.

Similarly, the centrality of the family and the moral imperatives of reciprocal social relationships modeled after graded kinship relations that define proper social conduct, all cherished in the reified Confucianism, further separate the Chinese and Thai moral universes (see also Bao 2005). The idea of saving for children or working hard to ensure family continuity, as most Chinese couples do, is considered unnecessary by Thais. This is also true of the reciprocal rights and obligations that permeate parent-child relationships in a typical Chinese family and are characterized as filial piety in Confucian teaching.

The positive image Yunnan Chinese hold of Thais is related to their peacefulness, honesty, and less complicated way of life. Mr. Huang, the former village *phuyaiban,* once complained to me about the betrayals and incessant infighting among Maehong villagers. He said that in his business dealings with the Thai, they tend to be more honest in keeping their promises than Yunnan Chinese and do not engage in backstabbing or clique-formation. As devoted Buddhists, Mr. Huang further elaborated, Thais always greet people with a smile and with a religious gesture, placing their hands together before them and bowing. In sharp contrast to the manipulative, hard-driven, and competitive Chinese, Mr. Huang added, the Thai are admired for an idyllic way of life that survives in Chinese culture only in "paradise lost" narratives.

Another villager, Mr. Zhang, who originally came from the Kogang area of Burma and settled in the village after marrying a village woman over 30 years ago, pointed out the differences between Thai and Chinese automobile drivers in a graphic way when he drove me to Chiang Mai in his pickup truck. He said, "You can easily pick out the Chinese in road traffic. Those who disobey traffic lights and always take short cuts while infringing on other people's road rights are definitely Chinese. The Thais are always relaxed, laid-back, and take things in their natural order. We Yunnan Chinese, however, are always in a rush, always trying to circumvent rules and regulations for personal advantage with no respect for others." But when I pressed Mr. Zhang for anything negative about Thais, he readily admitted that, compared to Yunnan Chinese, Thais are not as hard-working, nor are they willing to plan for the future. Because of those characteristics, he would not consider taking on a Thai business partner.

But no matter what the Chinese perceptions, Thais possess something many Yunnan Chinese, especially the new immigrants, do not have: an authentic document that permits them to live in Thailand legally. Villagers claim that, when King Bhumibol Adulyadej granted citizenship to the Yunnan Chinese in 1981, he originally intended to give them universal amnesty—that would include Nationalist military personnel, their families, and other Yunnan Chinese who lived in the village, such as caravan traders and their family. General Li, however, declined that offer. He feared that if all villagers received Thai citizenship, they would all move to Chiang Mai or Bangkok in search of a better life, and Maehong would be deserted. As a consequence, only military personnel and their immediate families received Thai citizenship. The others were granted various residential permits, but not citizenship.

With the continuous flow of illegal immigrants from China and Burma, the Maehong villagers' legal resident status has become extremely complex, having no fewer than eight categories in the Thai national classification system for residents: (1) full Thai citizens—those born in Thailand with at least one parent also a native-born Thai citizen—who can vote in elections and be elected; (2) naturalized citizens—such as the Nationalist soldiers who were granted citizenship in 1981—who can vote in elections but cannot become elected officials; (3) legal resident aliens—those Yunnan Chinese who received legal permits to reside in Maehong when their military contemporaries received citizenship—who cannot vote in elections but can travel anywhere in Thailand, and can apply to become Thai citizens eventually; (4) hill tribe members—legally Thai citizens since they are members of the Thai nation-state—due to their minority status, however, they cannot travel freely outside their tribal reservations; (5) war refugees—such as the Shan or Karen who escaped war in Burma to settle in Thailand—who can legally reside in the village and travel within Chaiprakarn Amphor; (6) certified hill tribe members—the Lisu, Shan, Lahu, or Akha from Burma whose full legal status is not established in Thailand, but due to their affinity with tribal groups on the Thai side of the border, they are accepted as quasi-Thai tribal members—who can stay only in hill areas and not in the village; (7) labor certificate holders—those who have established temporary residence in Thailand by virtue of employment; and lastly (8) overseas Chinese—those with legal documents issued by China or Taiwan who have obtained permission to reside and travel in Thailand legally, but only up to six months each time.

Illegal residents generally live in Maehong only temporarily. Some work on hill farms and live in farm shelters, earning about 3,000 baht per month (US$75, based on the 2003 exchange rate), as discussed in chapter 3. Others work in the village either as factory laborers or as domestic workers, earning about 2,000 baht per month. Because of their illegal status, they can only command wages below the market level. Better-educated illegal immigrants from China may find teaching jobs at the local Chinese school with better pay, discussed already in chapter 4. While living here, they learn the basics about Thai society, including the language, customs, employment opportunities, wage levels, and so on.

Once illegal immigrants save some money, they begin to explore the possibilities of obtaining legal status. This can be done in several ways. One is to bribe the county officials, especially Chaiprakarn Amphor's immigration officials, who can falsify a document to prove that the immigrant is from a certain hill area on the Burmese side of the border that entitles him or her to legal status as a tribal member. Another way to become a legal resident is to assume the name and identity of a deceased villager who held legal status. This method is expensive, since it takes permission from the deceased's family and the complicity of the local official who pretends not to have noticed the forgery. A third way to obtain legal status is to marry someone who is already legal. This also involves considerable costs and proper social relations. The last strategy used for obtaining legal status is waiting for the periodic amnesty from the Thai government; in these instances the village *phuyaiban* can appeal on behalf of those who have resided there. Mr. Huang claims that, in the last amnesty of 1999, he managed to obtain legal status for over one thousand people in Maehong.

Once they obtain legal documents, immigrants tend to move on to Chiang Mai or Bangkok, where jobs are easier to find and wages higher than in

Maehong. This type of sojourning seems to be the norm rather than the exception among the illegal residents in the village. They look upon themselves as transients rather than permanent residents of the village, Chaiprakarn Amphor, Chiang Mai Province, or even the Thai nation-state.

The highly contingent realities of life in Maehong have given rise to rather tenuous and ambivalent views about personal identity, ethnicity, and even nation-state affiliation (in the form of citizenship); this may be clearly seen among the offspring of the original village settlers. Youngsters born and raised in Maehong receive a complete Thai public education (from kindergarten up to grade 12) that underscores an idealized image of the Thai nation-state, centered on the Thai royal family (Keyes 1987; Van Esterik 2000; Wyatt 1982). They are perfectly fluent in Thai and are sufficiently familiar with the basic social norms and religious practices of the Thais to understand their parents' ethnocentric prejudices and misconceptions against the Thais—adding more resentment about the cultural baggage they have to carry in daily life.

Besides the formal Thai education, most of the village youth are corralled by their parents into attending the Chinese school, built at such enormous cost in money and energy by the first generation to preserve "Chinese culture." Many village children complain about this extra burden they must bear every day. They are also envious of the relatively carefree lifestyle their Thai friends enjoy. But rebellion against the Chinese school tends to be relatively muted. The village youth recognize the advantage the Chinese education provides by preparing them to compete in college entrance exams in Taiwan. Even those who decide not to pursue college education learn from life experience the advantages they have over their Thai contemporaries. Because of their bilingual (Chinese and Thai) or even trilingual (Chinese, Thai, and English) ability, village youth find ready employment in Bangkok's burgeoning international trade or in southern Thailand's tourist trade. The social advancement they experience, through monetary investment in education and through tiresome after-school lessons, is not only testimony to the importance of education; it also shows the limits of the easy-going local Thai culture.

In sum, village youths are better prepared to succeed in mainstream Thai society because they carry with them certain cultural traits their parents have inculcated in them that they reluctantly accept, such as respect for hard work and education, the will to plan for the future, and the competitive spirit that can propel them rapidly into the professional classes ahead of their Thai contemporaries. In other words, the young Yunnan Chinese diasporas, and perhaps this includes other Chinese immigrant groups in Thailand as well, are readily using elements of their Chinese ethnic identity—something they sometimes resent—to compete successfully in Thai society. Their acquisition of Chinese cultural practices and propensities have equipped them to occupy distinct urban, professional, and middle-class niches as wholesalers, retailers, bankers, educators, financial managers, manufacturers, real estate developers, international trade agents, and so on, in the open and permissive Thai society.

Once they occupy those positions, they readily reject cultural markers or practices associated with their Chinese past, along with its ethnic identity, to become Thai citizens. Looking at this process from the opposite point of view, we may suggest that it is perhaps because of such readily available opportunities for the young Chinese diasporas to successfully enter middle-class professions in Thailand that they are willing to compromise or relinquish their ethnic identity. The tradeoff between ethnicity and social class thus seems to distinguish the Yunnan Chinese diasporas from other Chinese diasporas in other Southeast Asia countries.

Some Maehong youth are ready to capitalize on this identity linkage to move into the global arena above and beyond the Thai nation-state. They extend their personal networks across regional, national, and even continental boundaries and rely heavily on the flexibility of being "Chinese." Many Maehong villagers have successfully obtained college admission (often with tuition subsidies) in Taiwan by recounting their commitment to the cause of anti-communism and their suffering as children of abandoned soldiers of the Nationalist government. It is estimated that over one thousand people from Maehong now permanently reside and work in Taiwan. Many who live in Taiwan have married locals and obtained legal resident status. Many of these still maintain their Thai citizenship for the sake of travel convenience and potential social and medical benefits.

The transnational ethnic networks that these villagers can participate in also provide the basis for joint economic endeavors, such as trade and investment. Processed local produce such as marinated ginger roots are exported to Japan through Taiwanese trade agents. Similarly, canned litchi, longan, bamboo shoots, and baby corn are sold to European and Southeast Asian markets through Chinese distributors in Hong Kong and Singapore. With improved living standards in China, there seems to be an insatiable appetite for dried longan as a cooking ingredient, and increasing numbers of mainland Chinese purchasing agents and commodity traders have been traveling to Maehong and nearby villages since 2000. Since most of these agents are new to the region, the introductions other ethnic Chinese can provide is essential; the presence of local Chinese smoothes the way for them to be properly received and share the latest information about regional market prices for specific local produce.

Finally, the transnational Chinese network also facilitates marriages when the parties seek partners of a similar ethnic background. Mr. Huang, the village's last *phuyaiban,* sent his daughter, Aihua, to the United States to study business management in the late 1990s. While in the U.S., Aihua stayed with a business acquaintance of her father's who matched her up with a second-generation Chinese American man. Mr. Huang cheerfully agreed to this marriage because he wanted his daughter to marry an ethnic Chinese, even though this young man is not Yunnan Chinese, since his grandparents came from Guangdong region, and despite the fact that he speaks no Chinese.

In summer 2005 another villager told me that his daughter was marrying a young man from Taiwan. The young man's parents have investments in Thailand and come often to inspect their manufacturing facilities. Both sets of parents have been active in Ciji Merit Association, the Buddhist Compassionate Society, in Bangkok; at one social gathering they met and were lamenting about their twenty-something unmarried children. They then decided that a match could be made and summoned the young man from Taiwan and arranged a date for them. Everything worked out well at the first meeting and six months later they were married.

The multilayered and multi-stranded Chinese ethnic network provides useful and reliable transnational channels that the Yunnan Chinese diasporas can readily tap into because all involving parties share similar cultural practices and values; when cooperative efforts come together, these help to expand their activities beyond nation-state boundaries. At a time when boundary-crossing migration, trade, and connections are essential for successful global competition, the Yunnan Chinese have in their cultural tool kit a ready-made instrument for attaining that success.

CONCLUSION

Maehong has evolved from a more or less homogenous community based on the military to a more diversified one as the generations have passed. Like other small communities, Maehong has seen the emergence of internal conflict due to different levels of wealth and the generation gap. Tensions have been further compounded by the Thai state's policy of extending civilian rule to the borderland, in addition to the prejudices ingrained in the ethnic hierarchy that the first generation Yunnan Chinese themselves observe.

For the Nationalist soldiers who came to northern Thailand from a multi-ethnic environment in China's southwest, the nation-state has remained a remote concept that merely provides the broadest political category in the Sinic world with little relevance in daily life—a view that seems to coincide with the Confucian ideals of universalism they so cherish (Harrell 1999). The modern concept of citizenship, whether based on a territorially defined political entity called the nation-state, or a set of biologically endowed genetic diacriticals obtained through descent, defies the traditional concept of "Chineseness" as a cultural marker determined by the degree of conformity to Confucian moral ideals.

The face-to-face encounter with the Thai nation-state, however, creates in the first generation Yunnan Chinese a sense of urgency to preserve and reproduce their own cultural heritage; in this process they have reified this social category "Chinese" as they have built a sustainable rural livelihood in this alien land. However, an unintended consequence of their active promotion of Chinese education, the extended family, popular rituals, and civilian morals, has been increased intergenerational conflict, internal dissension and a mass exodus of the youth. The disillusionment of these old soldiers with the Nationalists in Taiwan, their deep-seated suspicion of the Chinese Communist polity, and their contempt towards corrupt Thai bureaucracy has compelled them to embrace a de-territorialized notion of self and ethnic identity. Their tenacity in holding on to a petrified ideology, an anachronistic "Chineseness," may render them a true "lost" generation in search of a homestead to anchor their diminishing self-esteem and identity.

The offspring of the Yunnan Chinese settlers face an immediate social world that is predominantly Thai. They carry less cultural baggage than their parents regarding the supremacy of Confucian social values, and they are fully aware of the stereotypes, misconceptions and prejudices their parents hold towards Thais and other ethnic groups. For the younger generation, complete assimilation into mainstream Thai society is not only a viable but also an advantageous option. In this regard, they seemed to have confirmed Skinner's observation about Chinese immigrants in Thailand at large: after three generations, they are mostly assimilated into Thai society.

And yet many younger generation villagers are also fully aware that the cultural traits they obtained from their parents and from the Chinese education system have provided them with far more options than their Thai counterparts. Possibilities of social advancement are enhanced through education and employment in Taiwan, or in jobs in Bangkok or southern Thailand, where their multicultural experience and skills provide entrance into much higher level positions. Transnational ethnic networks also open up many opportunities in internationally oriented endeavors. As such, they see Chinese ethnicity as symbolic capital in a very different way than their parents: instead of preserving Chinese ethnicity as a marker of identity within the Thai nation-state, the younger generation is an active participant, willingly cultivating and manipulating this marker as instrument to social advancement and/or entry to transnational networks for trade, migration, and even marriage. With their pragmatic view of flexible citizenship, they are genuinely transnational citizens in this postmodern world.

Chapter 6

Conclusions

My last visit to Maehong was in summer of 2007, when I returned for a 10-day visit to collect information on Yunnan Chinese dietary culture and its change. It was during that time the village's Self Governing Committee (SGC), the governing body of Maehong since the 1960s but now a civilian organization, had its election of a new Director. The last election before this one was held in 1989 when Mr. Huang was handpicked by General Li to run the village through an allegedly "open" election. This time the election was orchestrated by the younger generation villagers who were controlling the village government and wanted to oust Mr. Huang, the current SGC Director, once and for all. This was the latest episode of Maehong's lingering intergenerational conflict that had been going on since 2000. Below are my field notes of that day's event on August 1, 2007:

> The village broadcast system announced at noon that the village Self Governing Committee will hold its annual meeting this evening at 7:00 sharp in the auditorium of the Chinese School. Mr. Zhou, the broadcast announcer and Maehong's mayor since 2002, couldn't conceal his joyous mood by urging all villagers to attend. He declared, "This represents a new chapter in our village history."
>
> I was having lunch at Mr. Liu's house when the broadcast was made. Upon hearing the announcement, Mr. Liu, the former president of the Chinese school board and a close ally of Mr. Huang, shook his head in dismay and said that it was unnecessary to humiliate Huang again in this open meeting since Huang had already conceded his defeat by turning over the Chinese school to Mr. Zhou in January, 2007. Mr. Liu said that Mr. Zhou still believed that there was enormous profit to be made from the Chinese school, and that was his motivation to completely expel Mr. Huang's remaining power that might intervene with his money making plans.

> I went to the school auditorium with Mr. Liu shortly before 7:00 pm to witness the last transfer of power. The auditorium was half empty. There were only a few women and children who came for this event. While women chitchat in small congregations, children ran around noisily. It was not until 7:30 that the auditorium was gradually filled.
>
> When the formal meeting began, Mr. Zhou went up to the podium and discussed recent village events. He then introduced the election commission that was appointed by him to identify the candidates for SGC. Mr. Liu grudgingly complained to me that all the seven commissioners were Zhou's drug-addict friends. After that, the commission chair introduced the sole candidate, Mr. Lin, who was selected by the commission. Mr. Lin came up to the podium and gave a brief speech about how he would use SGC to assist village affairs. Mr. Lin spoke with amazingly good Mandarin Chinese. Seeing my expressed puzzlement Mr. Liu told me that Mr. Lin spoke good Mandarin because he spent almost 20 years in Taiwan as a drug trafficker, and later went through drug rehabilitation programs there too.
>
> After Mr. Lin's short campaign speech, Mr. Zhou asked the audience to raise their hands if they would vote for Mr. Lin as SGC's new director. Fewer than half of the people raised their hands, and the remainders still seemed to be confused and uncertain as to what had happened. Seeing those raised hands, Mr. Zhou quickly declared that the overwhelming majority of the villagers had endorsed Mr. Lin's candidacy, and from then on Mr. Lin would replace Mr. Huang as the new Director of SGC. At that moment Mr. Zhou asked Mr. Huang to come up to the podium. Huang, wearing his usual long sleeve white shirt, went up stage and gave a moving speech in which he thanked all villagers for giving him the opportunity to serve them for the past 19 years. He also apologized for whatever mistakes he had made during his tenure. At the end of the talk, he presented the official seal wrapped in red satin to Mr. Lin. As he exited the auditorium, the impatient audience also began to disperse. This seemingly smooth transition of power belied the long struggle between the generations, signaled the arrival of a new era that would reorient Maehong to the mainstream Thai society, and cast a dark shadow over the Chinese diasporas' reproduction project initiated by the original settlers.

This latest development strikes me both intellectually and personally. Given my particular life experience as a "Chinese," I am to a certain degree empathetic of what the Yunnan Chinese diasporas, both young and old, are engaging. I was born to a Shanghai mother and an eastern Cantonese father (i.e., Chaozhou, or more popularly known as Teochiu) in Sichuan, China. For this reason, my first language was Sichuanese supplemented by my mother's Shanghai language. At the age of 4, my family moved to Hainan Island and later to Hong Kong where my third language became Cantonese. After 1951, I moved to central Taiwan where Hokkianese was my daily language with classmates and local folks and Mandarin Chinese the official language. Since 1970, and for the next 35 years, my daily and professional language had changed to English when I attended graduate school at Michigan State University and, later, my teaching career in anthropology began in Iowa. Looking at my own life course, I can handle several distinct languages, shifting from one to the next with ease, and yet I am never able to speak the language of my father's "hometown"—Chaozhou. Similarly, I have moved across different localities in China, Taiwan, and the U.S. and established residence for various durations there, and yet really belong to none. This life experience, coupled with my anthropological insights, made me ponder the meaning of "hometown" as well as what type of personal identity can be constructed if not couched in the straightjacket of the nation-state.

Some people might be more concerned about constructing this type of self-identity—ascribed, learned, or fantasized—and may spend a considerable amount of time pursuing a solution for it. My exploration of the Yunnan Chinese diasporas in Maehong of Northern Thailand is particularly exemplary of this life inquiry. For the first generation settlers, they pushed hard to realize their "Chineseness" through strategically maneuvering it in this alien land: cunningly deploying their stock of practical knowledge in agricultural production and cultivating business networks with other overseas Chinese. Furthermore, given their explicit anti-communist ideological commitment in the global Cold War era, they succeeded in ensuring their survival by attracting the support of the anti-communist coalition in Southeast Asia (e.g., Thailand, the U.S., and Nationalist Taiwan), followed by tapping into the illicit drug production and trade and then converting themselves into successful tropical fruit producers in the hills. From this valorized self-identity, they have, as discussed in chapter 3, built a seemingly viable agrarian community characterized by concern with environmental conservation, economic viability and, less certainly, sociocultural renewal—all yardsticks of agricultural sustainability.

The pursuit of the life answer by the original settlers in Maehong, however, has presented a challenge to my anthropological thinking because their effort seemed to have generated mostly resentment and defiance from their children. Is this merely a reflection of normal intergenerational conflict virtually all society must cope with? How do I assess the success or failure of reproducing Chinese diasporas in Maehong across space and time? Or is this a relevant issue at all? To get to the bottom of my inquiry, I have to cast this miniscule ethnographic locale against the global whirlpool that is extending its all-embracing reaches to this remote corner of Thailand's northern hill regions.

When disentangling this space-time compression embedded in the production and reproduction of Yunnan Chinese diasporas in Maehong, I wish to separate two layers of analyses in light of my foregoing ethnographic presentation. The first one involves the practices deployed by the original migrants in constructing and sustaining this transplanted community through intertwining an imagined Chinese identity with contested cultural practices. Although pragmatic in certain senses, I want to explain how these cultural practices also engender a social order characterized by rigidity, resentment and ultimately self-negation over time.

The second analysis I wish to cover in the conclusion deals with the theoretical implications this study entails in our current understanding of Chinese diasporas in the context of globalization. Globalization as the grand narrative of the 21st century has engendered many changes: the increase of cross-border trade (i.e., the neo-liberal, deregulatory economic policy minimizes tariffs), the realignment of the nation-states as containers of social process (i.e., the weakening of the nation-state as the regulator of domestic and international affairs), the production and reproduction of global diasporas (i.e., transnational subjects who exhibit high levels of fluidity in geographic movements and multiple identities), and the rise of cross-border connectivity and new global aspiration and social entity.[1] Issues raised by contemporary theorists can be teased out based on the Maehong data.

THE PRACTICES OF DIASPORAS CONSTRUCTION IN MAEHONG

The particularities of the Maehong Chinese diasporic community lie in its ambivalent position regarding an imagined "cultural China" and Chinese ethnicity. When leaving China in the late 1940s, the Yunnan Chinese believed that their cultural heritage was in jeopardy because the encroaching Chinese Communists represented a devious international conspiracy to destroy Chinese culture. For the first generation settlers, the harshness and struggle they endured in the Golden Triangle, which subsequently became a part of their collective memories, as well as their involvements in the illicit drug trade, were sustained by the idea that they were the guardians and transmitters of a genuine Chinese cultural heritage.

These shared experiences and practices are the decisive factors that enabled the Yunnan Chinese to sustain their collective "social suffering," as defined by Kleinman et al., which "results from what political, economic, and institutional power does to people and, reciprocally, from how these forms of power themselves influence responses to social problems" (Kleinman, et al., 1997). In Maehong, the first generation Yunnan Chinese endured their collective suffering initially as displaced war refugees driven out of China and, later, as a minority enclave in the hills of northern Thailand. It is in this shared historical experience that we find how the first generation diasporas and their collective memories of social suffering metamorphosed into something even bigger than their mere survival: "Historical memories of suffering . . . have present uses . . . to authorize nationalism or class and ethnic resistance" (Kleinman et al. 1997: xi).

It is with this extreme form of nationalism or hardened ethnic resistance that the original Yunnan Chinese identified the idealized Confucian culture as the anchoring point of their "longing in belonging" (Ilcan 2002). To them, citizenship or nationhood is significant only to the extent that it is associated with or reinforces their recognized traditional cultural markers based on a reified Confucianism. Most of them accept Thai citizenship only for the sake of long-term residence in Thailand, but their ethnic identity sticks to "cultural China," rather than the territory-bounded Chinese polity, either in China or Taiwan.

Their initial success has caused this reified culture to acquire the distinct characteristics of infallibility which could not be challenged. Living with nostalgia, the original settlers in Maehong, the victims of so much social suffering, could nevertheless turn themselves around by establishing a social order to inflict misery upon the weak, the less fortunate, and the discriminated. Their efforts, grounded in both rigid military hierarchy and calcified Confucian moral teaching, replicate an idealized Chinese society and have resulted in inflexible, morbid, and anachronistic attitudes and practices, a process I call "cultural petrification."

By petrification, I am referring to a development trajectory in that the inherited cultural practices have acquired a formalistic characteristic or authority which stifles creativity, silences dissension, and resolves internal conflicts through rigid interpretation of the orthodoxy—the terminal end of cultural change. Previous research has provided examples that identified such a convoluted cultural calcification process. For instance, in Chinese history, historian Mark Elvin (1977) coined the term "High Level Equilibrium Trap" to describe the general lack of self-innovative capability of Chinese culture after the glorious Song Dynasty (circa CE 960–1278). The self-valorization of cultural accomplishment during this time period had produced and fortified a sense of superiority that, in turn, discouraged innovative changes.[2]

Maehong's cultural petrification is similar to the above-mentioned non-progressive, calcifying condition. It is as if their lived universe is frozen in time. Confucian morals are exhorted in official public gatherings, in school textbooks, and in popular ritual performances. In this moral universe, the "superiors" (i.e., the military commanders, the factory owners, the fathers, and the husbands) are supposed to take care of their "subordinates" (the rank-and-file soldiers, the floor workers, the sons and daughters, and the wives), and the former often play up the rhetoric in public and demand loyalty and submission from the latter in return. And yet whenever given the opportunity, the superiors often take advantage of the subordinates by skimming their payments, forcing them into illicit activities, or simply selling the daughters into prostitution to support their drug addiction.

Hence in the perceived social hierarchy in Maehong's everyday life, we find mostly hypocrisy and pretentiousness emanated from top to bottom, and in reverse direction too. Those who have power or authority and are in a position to dispense care or material benefit to the subordinates would typically declare, in self-congratulatory tones, the amount of sacrifices they have made for the latter. They would then complain that their good deeds are never fully acknowledged or appreciated by the inferiors in return. In the opposite direction of this social intercourse, however, we find mainly disgruntlement and distrust of the inferiors against the superiors—not only expressed grudgingly in private but also openly in public discourses.

This type of mutual contempt or lack of trust permeates all corners of Maehong's social life. Virtually all public figures in Maehong have been accused of taking graft or engaging in dishonest conduct at the expense of the general public. General Li, the paternal figure of the entire Third Army, for example, was fingered to be the one who benefited the most from the illicit drug trade. This type of accusation against official corruption and personal misconduct, although not to the extent of General Li, has been lodged against virtually all other public figures such as high ranking military officers immediately below General Li, the current and previous heads of the village's Self Governing Committee, the current and previous school principals, and so on and so forth.

Perhaps the problem in Maehong is more severe than the other Yunnan Chinese communities in northern Thailand simply because of its firmer military control, and then it is also the most rebellious—defiant youngsters either abandoned the community entirely or staged their attacks on the establishment by invoking Thai authorities. The social milieu in Maehong reached exactly such a level of morbidity (of the original diasporas) and resistance (among their offspring) at the turn of the century when the younger generations rose up to challenge the ruling oligarchy and succeeded in unseating them from the village power hierarchy.

The irony here is: even though the reproduction of an imagined culture has enabled the first generation Yunnan Chinese to establish a proper rural livelihood in Maehong, that culture in fact distinguishes them from other minority groups in the Golden Triangle as well as from other Chinese communities in China, Taiwan, and elsewhere in Thailand. The tenacity they possessed in reconstructing or sustaining an imagined culture long past has produced a social anachronism that proves to be formalistic, inflexible, and lacking in innovation. They live a life which is frozen in time and space of a reified Chineseness.

The second- and third-generation Yunnan Chinese in Maehong have acquired a full-fledged Thai education and are comfortable with the Thai way of life. To them, the cultural markers that their parents defined are something of the remote past. However, they have strategically exploited aspects of the Chinese cultural heritage acquired from either their parents or the local Chinese school—such as language capabilities, monetary acumen, an inclination to hard work, and emphasis on interpersonal relations—and converted them into useful sociocultural capital for rapid and often successful mobility into the professional classes. This ethnicity or cultural identity is instrumental to acquiring a good life in their chosen nation-state—Thailand.

Those not content with the options provided by the Thai nation-state look beyond the national borders for resettlement opportunities in other places, such as Taiwan, Singapore, Australia, Europe, and North America. Thus, a person's decision to become either a transitional sojourner or a long-term citizen of a nation-state is contingent on many situational considerations that defy simple labeling. Their concerns about citizenship and national identity seem to be transitional and conditional, contingent upon individual calculations of alternatives in pursuance of a good and properly moral life, no longer defined by Confucian ideals.

Such a development fits well to what Stuart Hall describes as a new form of self-identity in the post-modern world: "the identities become detached—disembedded—from specific times, places, histories, and traditions, and appear 'free-floating'" (1992: 302–303). Hall also points out, "nationalism and ethnicity were archaic forms of attachments—the sort of thing which would be melted away by the revolutionizing force of modernity" (Hall 1992: 314). In the hills of northern Thailand, as the extending arms of globalization gradually peel away the protective shields of a reified Chineseness, what we are witnessing is the formation of new generations of diasporic Chinese who will take on new social identities and global perspectives that are fundamentally different from their forebears.

CHINESE DIASPORAS AND GLOBALIZATION

Juxtaposing the past and the present, how can we position the Yunnan Chinese in Maehong in the study of Chinese diasporas and globalization? I am hesitant to project my findings in Maehong to Chinese overseas in Southeast Asia in general as observed by Ong (1999). The Yunnan Chinese diasporas were facing an overall environment which was significantly different from other Chinese overseas communities in several regards. The first one is related to the nature of Thai society. As a flexible and highly tolerant society, Thailand can be regarded as an ideal plural society. Expatriates can easily find their personal niche and be accepted in the Thai social milieu. This is certainly a welcome alternative to the rigid, stifling, and pain-inflicting social environment that the original Yunnan Chinese established in Maehong.

The second reason that Maehong's younger generations seemed to be eager to relinquish their diasporic label, I speculate, is associated with the infamy of the Yunnan Chinese's earlier drug production and trade. Even in a highly tolerant society such as Thailand, the two decades of opium production and its refinement into heroin and trade across international borders from the 1960s through the 1980s have made the term "Yunnan Chinese" synonymous with illegal conduct among the Thai public. When the options became available to "passing" as other national subjects, the younger generation diasporas readily embraced them. Besides, the increasing opportunities for the new generation Yunnan Chinese to engage in secondary migration to other countries provide still other alternatives if they are not content with being assimilated into mainstream Thai society.

At the end of this chapter, I wish to take my analytic terrains in Maehong to another level, addressing theoretical issues related to forming diasporas, globalization, transnationalism, and the realignment of the nation-state in the context of globalization. First, as a group of border-crossing refugees, the Yunnan Chinese can be regarded as a lived example of the primordial transnational subjects who seek to establish a post-national existence. In this context, we may ask questions such as: How have their de-territorial propensity of exiting China and their construction of transnational networks contributed to the de-centering of the state, as seen in the weakening of the holdings of the nation-state, an issue that globalization theorists have been debating (Appadurai 1996; Lewellen 2002; Ong 1999; Parrenas and Siu 2007; Walker 1999)? When the first generation diasporas in Maehong deployed their cultural knowledge and social capital to sustain their presence in northern Thailand's multiethnic mosaic, can we regard this as the embodiment of a supra-national entity that transcends the traditional nation-state which juxtaposes people, culture, language, territory, sovereignty, and citizenship in a single alignment? Conversely, when many of the second or third generation offspring of the original migrants readily abandoned the Chinese identity that their forebears so cherished and eagerly merged into mainstream Thai society or other political entities, can we consider this to be another sign of the demise of the nation-state as the current globalization process accelerates?

Second, at a time of increasing global flows of information, goods, capital, and even humans crisscrossing nation-state boundaries, it has been suggested that the fluidity of the postmodern livelihood will inevitably lead to the weakening of the pre-modern modes of naturalized self-identity, be it national, class, ethnic, linguistic, or gender (Appadurai 1996; Horstmann and Wadley 2006; Ong 1999; Parrenas and Siu 2007; Sassen 2007; Walker 1999; Yeoh, Charney and Kiong 2003). To what extent have the Yunnan Chinese diasporas fit this characterization and, in the same vein, what type of fluid and multiple post-national order they have acquired as replacement? Have they developed a transnational forum which Appadurai (1996: 22) calls "diasporic public spheres" as a substitute for their primordial identity? How do they maintain their diasporic connectivity through the "everyday practices of sociality, collective memory, economic exchange, and the work of cultural imagination and production . . . " (Parrenas and Siu 2007: 2)?

Based on my data in Maehong, it seems that some scholars' suggestion about the demise of the nation-state is premature—a point generally concurred with by other researchers (Lewellen 2002; Ong 1999; Walker 1999). The first generation Maehong diasporas anchored their nation-state identity with the imagined "cultural China" (Tu 1994). For the second and third generation diasporas, once they rejected the petrified and reified cultural China, they turned quickly to the trapdoors of other nation-states—Thailand, Taiwan, or any other countries that promised opportunities for upward mobility. For the purpose of personal convenience or contingent needs, they may hold multiple identities or even multiple passports. But the presence of the nation-state as the "container of social process" (Sassen 2007) is never questioned or challenged. The Yunnan Chinese diasporas may be transnational subjects *par excellence,* but their lived experiences predicate on the existence of the nation-state.

Similarly, what do we make of the issue that some globalization theorists suggested, namely about the possibility of constructing new transnational entities, either the "diasporic public spheres" which initiate the post-national political order (Appadurai 1996) or "overseas Chinese fraternal networks" that undermine Eurocentric capitalism (Ong 1999)? My findings in Maehong do not lend conclusive evidence to either support or reject these theses. The Yunnan Chinese diasporas, including the original settlers and their offspring, do maintain active connections across nation-state boundaries. They seek information about opportunities in trade, employment, education, or even marriage from kin, friends, and co-workers in other localities, in Thailand or beyond. Actively cultivating interpersonal relationships and converting them into effective sociocultural capital have always been a part of Chinese social life; in this regard they seem to share similar cultural traits with other Chinese diasporic communities elsewhere in the world.

Has this type of transnational connectivity ever been elevated to become a new kind of consciousness that challenges either the existing nation-state or the hegemonic Western capitalism? I am rather doubtful about such propositions. The first generation Maehong diasporas might have harbored a strong anti-communist ideology and hence entertained certain global visions about local events. For the majority of the Maehong's second and third generations residents, however, they are more concerning about immediate personal needs, such as completing a proper education, finding a job, establishing financial self-sufficiency, building a family, and so forth. In other words, they may have developed new a personal subjectivity that deviates from that of their parents or grandparents, and yet such subjectivity is rather mundane and does not lend them a vision beyond the close personal confine. To acquire a more satisfactory life, the younger generation Maehong diasporas are ready to shake off their Chineseness for another nation-state label. In this sense they not only render their parents' or grandparents' diasporic production project a failure, but also disappoint the globalization theorists for their inability to promote a new transnational social order.

Thus my story of reproducing Chinese diasporas in Maehong is still inconclusive because the contesting forces of the global and the local are exerting their enormous power to shape this process and its outcomes. The original settlers built their cultural fortress in Thailand's northern wilderness, which seemed to be successful initially when a replica of the Confucian social order was established. The self-valorized cultural heritage, however, bred its own contradiction and conflict when cast against the historically kinder society of Thailand. Globalization provided the readily available alternatives with which Maehong's new generation diasporas discarded their petrified cultural garment for other naturalized self-identities, but their options seemed to be far more diverse than any theoretical models have predicted. Globalization is indeed the grand narrative of our current era, but the global citizens emerging from Maehong still anchor their primary identity in the nation-state(s) in which they maintain socially recognizable status.

NOTES

1. For instance, Ong (1999: 18) observed among Chinese diasporas in general who entailed "both national and transnational forms of nationalism that not only reject Western hegemony but seek, in pan religious civilizational discourses, to promote the ascendancy of the East." They have further established "new modes of subject making and the new kinds of valorized subjectivity" (Ong 1999: 19).

2. Similarly, Geertz, in his study of the convoluted agrarian society in Dutch colonial Java, used the term "cultural involution" to describe the stifled development process. This Javanese involution, as a counter-evolutionary and counter progressive process, is characterized by increased population pressure on limited agrarian resources, the extractive nature of the Dutch colonial regime that prevented technological change, and the rural populace's reliance on increasingly intricate social and spiritual rituals to spread out shared poverty (Geertz 1963).

References

Anderson, Edward. 1993. *Plants and people of the Golden Triangle: Ethnobotany of the Hill Tribes of Northern Thailand.* Portland, OR: Dioscorides Press.

Appadurai, Arjun. 1996. *Modernity at large: Cultural dimensions of globalization.* Minneapolis: University of Minnesota Press.

Auansakul, Panee. 1995. Chinese traders and Thai groups in the rice business. In *The Ethnic Chinese of Thailand,* eds. Chan and Tong, 23: 29–42. A special issue of *Southeast Asian Journal of Social Science.*

Bao, Jiemin. 1995. Sino-Thai ethnic identity: Married daughters of China, and daughters-in-law of Thailand. In *The Ethnic Chinese of Thailand,* eds. Chan and Tong, 23: 57–77. A special issue of *Southeast Asian Journal of Social Science.*

Bao, Jiemin. 2005. *Marital acts: Gender, sexuality, and identity among the Chinese Thai diaspora.* Honolulu: University of Hawaii Press.

Bates, D. G. 2001. *Human adaptative strategies: Ecology, culture, and politics.* Boston: Allyn and Bacon.

Bo, Yang (Deng Kebao). 1961. *Yiyu* (Alien Lands). Taipei: Hsing–kuang Publications.

Bo, Yang (Deng Kebao). 1982. *Jinsanjiao, bianqu, huangcheng* (*Golden Triangle, borderland, and deserted city*). Taipei: China Times Publication.

Casey, Edward S. 1996. How to get from space to place in a fairly short stretch of time: Phenomenological prolegomena. In *Senses of Place,* eds. S. Feld and K. H. Basso. Santa Fe, NM: School of American Research Press.

Chamratrithirong, Aphichat. 1984. Loosely-structured Thailand: the Evidence from marriage culture. In *Perspectives on the Thai Marriage,* ed. Aphichat Chamratrithirong, Bangkok: Institute for Population and Social Research, Mahidol University Publications.

Chan Kwok Bun and Tong Chee Kiong. 1993. Rethinking assimilation and ethnicity: The Chinese in Thailand. *International Migration Review* 27: 140–168.

Chan Kwok Bun and Tong Chee Kiong. 1995. Modeling culture contact and Chinese ethnicity in Thailand. In *The Ethnic Chinese of Thailand*, eds. Chan and Tong, 23: 1–12. A special issue of *Southeast Asian Journal of Social Science*.

Chang, Wen-Chin. 2001. From war refugees to immigrants: The case of the KMT Yunnanese Chinese in northern Thailand. *International Migration Review* 35: 1086–1105.

Chang, Wen-Chin. 2002. Identification of leadership among the KMT Yunnanese in northern Thailand. *Journal of Southeast Asian Studies* 33: 123–146.

Chang, Wen-Chin. 2005. Invisible warriors: The migrant Yunnanese women in northern Thailand. KOLOR: *Journal on Moving Communities* 5: 49–70.

Chuang,Yin-chang. 1985. Family structure and reproductive patterns in a Taiwanese fishing village. In *The Chinese Family and Its Ritual Behavior*, eds. Hsieh and Chuang. Taipei: Institute of Ethnology. Academia Sinica Publications.

Cohen, Myron. 1976. *House united, house divided: The Chinese family in Taiwan.* New York: Columbia University Press.

Coughlin, Richard. 1960. *Double identity: The Chinese in modern Thailand.* Hong Kong: Hong Kong University Press.

Cowell, Adrian. 2005. Opium anarchy in the Shan state of Burma. In *Trouble in the triangle: Opium and conflict in Burma*, eds. Martin Jelsma, Tom Kramer, and Pietje Vervest. Chiang Mai, Thailand: Silkworm Books.

Croll, Elisabeth. 1983. *Chinese women since Mao.* London: Zed Books.

De Bary, William. 1988. *East Asian civilizations: A dialogue in five stages.* Cambridge: Harvard University Press.

Demmers, Jolle. 2007. Wars and diasporas: Suggestions for research and policy. *Journal of Peace Conflict & Development Issue* 11: 1–26.

Deng, Xian. 2000. *Liulang jinsanjiao* (*Drifting in the Golden Triangle*). Beijing: People's Literature Publications.

Ding, Yuling. 2005. Economic activities and the construction of gender status among the Xunpu women in Fujian. In *Southern Fujian: Reproduction of traditions in post-Mao China,* ed. Tan Chee-Beng. Hong Kong: The Chinese University Press.

Duara, Prasenjit. 1990. *Culture, power and the state: Rural North China, 1900–1942.* Stanford: Stanford University Press.

Easman, Milton. 1986. The Chinese diaspora in Southeast Asia. In *Modern diaspora in international politics,* ed. Gabriel Sheffer. London: Croom Helm.

Edelman, Marc, and Angelique Haugerud. 2004. Introduction: The anthropology of development and globalization. In *The anthropology of development: From classical political economy to contemporary neoliberalism,* eds. Edelman and Haugerud. Malden, MA, Oxford, U.K., and Carlton, Victoria, Australia: Blackwell Publishing.

Elvin, Mark. 1977. *The pattern of the Chinese past.* Stanford: Stanford University Press.

Embree, John. 1969. Thailand—A loosely structured social system. In *Loosely structured social systems: Thailand in comparative perspective,* ed. Hans-Dieter

Evers. New Haven: Yale University Southeast Asia Studies, Cultural Report Series No. 17.

Escobar, Arturo. 1994. *Encountering development: The making and unmaking of the Third World.* Princeton, NJ: Princeton University Press.

Feingold, David. 2000. The hell of good intentions: Some preliminary thoughts on opium in the political ecology of the trade in girls and women. In *When China meets Southeast Asia: Social and cultural change in the border regions,* eds. G. Evans, C. Hutton, and K. K. Eng. New York: St. Martin's Press.

Ferguson, James. 1994a. *The anti-politics machine: "Development," depolitization, and bureaucratic power in Lesotho.* Minneapolis: University of Minnesota Press.

Ferguson, James. 1994b. Anthropology and its evil twin: "Development" in the constitution of a discipline. In *International development and the social sciences: Essays on the history and politics of knowledge,* eds. F. Cooper and R. Packard. Berkeley: University of California Press.

Feuchtwang, Stephan. 1992. *The imperial metaphor: Popular religion in China.* London: Routledge.

Forbes, Andrew, and David Henley. 1997. *The Haw: Traders of the Golden Triangle.* Bangkok, Thailand: Amarin Printing and Publishing Company.

Fox, Jefferson. 2001. Land use and cover change in montane mainland Southeast Asia. In *Land use and cover change,* eds. R. B. Singh, J. Fox, and Y. Himiyana. Enfield: Science Publishers, Inc.

Francis, Charles. 1990. Sustainable agriculture: Myths and realities. *Journal of Sustainable Agriculture* 1: 97–106.

Fricker, Alan. 1998. Measuring up to sustainability. *Futures* 30: 367–375.

Geddes, W. R. 1976. *Migrants of the mountains: The cultural ecology of the Blue Miao (Hmong Njua) of Thailand.* Oxford: Clarendon Press.

Geddes, W. R. 1983. Research and the tribal research centre. In *Highlanders of Thailand,* eds. John McKinnon and Wanat Bhruksasri, 3–12. Oxford: Oxford University Press.

Geertz, Clifford. 1963. *Agricultural involution.* Berkeley: University of California Press.

Geertz, Clifford. 1966. Religion as a cultural system. In *Anthropological approaches to the study of religion,* ed. M. Banton. London: Tavistock Publications.

Giddens, Anthony. 1991. *Modernity and self-identity: Self and society in the late modern age.* Oxford: Polity Press.

Gold, Mary. 1999. *Sustainable Agriculture: Definitions and Terms.* Baltimore: U.S. Department of Agriculture.

Hall, Stuart. 1992. Question of cultural identity. In *Modernity and its futures,* eds. S. Hall, D. Held, and T. McGrew. Cambridge: Polity.

Hall, Stuart. 1996. Introduction: Who needs "identity"? In *Questions of cultural identity,* eds. S. Hall and P. D. Gay. London: Sage Publications.

Han, Ju Hui Judy. 2005. *Missionary destination and diasporic destiny: Spatiality of Korean/American evangelism and the cell church.* Institute for the Study of Social Change, University of California, Berkeley, Working Papers.

Hansen, Art. 1994. The illusion of local sustainability and self-sufficiency: Famine in a border area of Northwestern Zambia. *Human Organization* 53: 11-20.

Harper, Janice. 2002. *Endangered species: Health, illness and death among Madagascar's People of the Forest.* Durham, NC: Carolina Academic Press.

Harrell, Stevan. 1985. Why do the Chinese work so hard? Reflections on an entrepreneurial ethic. *Modern China* 11: 203–226.

Harrell, Stevan. 1999. The role of the periphery in Chinese nationalism. In *Imagining China: Regional division and national unity,* eds. S. Huang and C. Hsu. Taipei: Institute of Ethnology, Academia.

Hatfield, Jerry, and Dennis Keeney. 1994. Challenges for the 21st century. In *Sustainable agriculture systems,* eds. J. L. Hatfield and D. L. Karlen. Boca Raton: Lewis Publishers.

Heidhues, Mary Somers. 1996. Chinese settlements in rural Southeast Asia: Unwritten histories. In *Sojourners and settlers: Histories of Southeast Asian and the Chinese,* ed. Anthony Reid, 164–182. St. Leonards, Australia: Allen & Unwin.

Helmore, Kristin, and Naresh Singh. 2001. *Sustainable livelihood: Building on the wealth of the poor.* Bloomfield, CT: Kumarian Press.

Horstmann, Alexander, and Reed L. Wadley. 2006. *Centering the margin: Agency and narrative in Southeast Asian borderlands.* New York: Berghahn Books.

Hsieh, Jih-chang. 1985. Meal rotation. In *The Chinese family and its ritual behavior,* eds. Hsieh and Chuang. Taipei: Institute of Ethnology, Academia Sinica Publications.

Hsieh, Shih-chung. 1997. *Guozu guojia de jiango, fanchou, yu zhibian: zhunghua-minguo lujun di jiushihsan shi de yunnan, mian, tai, taiwan pan shih-chi* (The construction, category, and metamorphosis of the nation-state: the half century of the 93rd division of the Nationalist Army in Yunnan, Burma, and Thailand). *Bulletin of the Department of Anthropology, National Taiwan University* 52: 43–68.

Huang, Shu-min. 1991. Re-examining the ideal, extended family in Chinese peasant society: Evidence from a Fujian village. *The Australian Journal of Chinese Affairs* 27: 25–38.

Huang, Shu-min. 2001. Economic culture and moral assumptions in a Chinese village in Fujian. *Asian Anthropology* 1: 59–86.

Huang, Shu-min. 2006. A Chinese Diasporic community in northern Thailand: Contested political loyalty and shifting ethnicity. In *Chinese transnational networks,* ed. C. B. Tan. London: Routledge Publications.

Husband, Charles. 1998. Globalization, media infrastructure and identities in a diasporic community. *The Public* 5: 19–33.

Ikerd, John E. 1992. Sustainable agriculture, A matter of people. *Issues in sustainable agriculture.* Center for Sustainable Agricultural Systems, University of Missouri, Columbia.

Ilcan, Susan. 2002. *Longing in belonging: The cultural politics of settlement.* Westport, CT: Praeger Publications.

Inda, Johathan Xavier, and Renato Rosaldo. 2002. Introduction: A world in motion. In *The anthropology of globalization: A reader,* eds. Inda and Rosaldo. Malden, MA: Blackwell Publishers.

Johnson, Kay J. 1983. *Women, the family, and peasant revolution in China.* Chicago: University of Chicago Press.

Keyes, Charles. 1987. *Thailand: Buddhist kingdom as modern nation-state.* Boulder: Westview Press.

Kleinman, Arthur, Veena Das, and Margaret Lock. 1996. Introduction. In *Social suffering,* eds. Kleinman, Arthur, Veena Das, and Margaret Lock. Berkeley: University of California Press.

Knox, M. L. 1989. Nation an island. *Sierra,* May/June, 78–84.

Kung, Lydia. 1983. *Factory women in Taiwan.* Ann Arbor, MI: UMI Research Press.

Kunstadter, Peter. 1983. Highland populations in Northern Thailand. In *Highlanders of Thailand,* eds. John McKinnon and Wanat Bhruksasri, 15–45. Oxford: Oxford University Press.

LeBar, Frank M., Gerald C. Hickey, and John Musgrave. 1964. *Ethnic groups of mainland Southeast Asia.* New Haven: Human Relations Area Files Press.

Levin, Simon A. 1992. The problem of pattern and scale in ecology: The Robert H. MacArthur Award Lecture. *Ecology* 73: 1943–1967.

Lewellen, Ted. C. 2000. *Anthropology of globalization: Cultural anthropology enters the 21st century.* Westport: Bergin & Garvey.

Lewis, Paul, and Elaine Lewis. 1984. *People of the Golden Triangle: Six tribes in Thailand.* London: Thames & Hudson, Ltd.

Li, Jian. 2001a. Cogon-thatched cottages and iron sheet-roofed houses: Development in a Yao Mountain village in northern Thailand. *Culture and Agriculture* 23: 24–40.

Li, Jian. 2001b. Development and tribal agricultural economy in a Yao Mountain village in Northern Thailand. *Human Organization* 60: 80–94.

Li, Yi-yuan. 1985. On conflicting interpretations of Chinese family rituals. In *The Chinese family and its ritual behavior,* eds. Hsieh and Chuang. Taipei: Institute of Ethnology, Academia Sinica Publications.

Li, Yi-yuan. 1986. 中國家族及其儀式：若干觀念的檢討 (Chinese family and rituals: Reassessment of some concepts.) In 中央研究院民族學研究所集刊 59: 47–61.

Li, Yi-yuan. 1993. 人類的視野 (Visions of anthropology). Shanghai: Wenyi Publications.

Lien, R. Z. 2005. 姓氏與祖先：雲南洱海地區社會階序的形成 (Surnames and ancestors: The formation of social hierarchy in the Erhai region of Yunnan). *歷史人類學* 4: 1–36.

Lin, Nan. 2000. *Social Capital: A theory of social structure and action.* Cambridge: Cambridge University Press.

Liu, Ts'ui-Jung. 1992. *明清時期家族人口與社會經濟變遷* (*Lineage population and socioeconomic changes in the Ming-Ch'ing periods*). Taipei: Institute of Economics, Academia Sinica.

Maxwell-Hill, Ann. 1983. The Yunnanese: Overland Chinese in northern Thailand. In *Highlanders of Thailand,* eds. John McKinnon and Wanat Bhruksasri, 123–133. Oxford: Oxford University Press.

Maxwell-Hill, Ann. 1998. *Merchants and migrants: Ethnicity and trade among Yunnanese Chinese in Southeast Asia.* New Haven: Yale University Southeast Asia Studies, Monograph 47.

McCoy, Alfred W. 1972. *The politics of heroin in Southeast Asia.* With Cathleen B. Read and Leonard P. Adams II. New York: Harper & Row.

Mote, F. W. 1967. The rural "Haw" (Yunnanese Chinese) of northern Thailand. In *Southeast Asian tribes, minorities, and nations,* ed. Peter Kunstadter, 487–524. Princeton, NJ: Princeton University Press.

Nee, Victor. 1983. Chinese peasant familism. In *Chinese rural development: The great transformation,* ed. William Parish. Armonk, NY: M. E. Sharpe.

Ong, Aihwa. 1999. *Flexible citizens: The cultural logics of transnationality.* Durham, NC: Duke University Press.

Ongprasert, S., and F. Turkelboom. 1997. 20 years of alley cropping research and extension in the sSlopes of Northern Thailand. Handout paper for the Training Course on Sustainable Highland Agriculture Development. Faculty of Agriculture, Chiang Mai University.

Parrenas, Rhacel S., and Lok C.D. Siu, eds. 2007. *Asian diasporas: New formations, new conceptions.* Stanford: Stanford University Press.

Panomtarinichigul, M., C. Sukkasem, S. Peukrai, M. Fullen, T. Hocking, and D. Mitchell. 2000. Comparative evaluation of cultural practices to conserve soil and water on the highland slope in northern Thailand: Multidisciplinary approaches to soil conservation strategies. *Proceedings of International Symposium ESSC, DBG, ZALF.* Muncheberg, Germany.

Panomtarinichigul, M., and Shu-min Huang. 2007. Implications of Yunnan Chinese land-use practices for sustainable hill farming in northern Thailand. *Taiwan Journal of Anthropology* 5: 123–147.

Pieterse, Jan N. 2001. *Development theory: Deconstruction/reconstruction.* London: Sage Publications.

Pongsapich, Amara. 1995. Chinese settlers and their role in modern Thailand. In *The Ethnic Chinese of Thailand,* eds. Chan and Tong, 23: 13–28. A special issue of *Southeast Asian Journal of Social Science.*

Purcell, Victor. 1951. *The Chinese in Southeast Asia.* London, Kuala Lumpur, and Hong Kong: Oxford University Press.

Rakariyatham, Pong-In. 2004. Land use and land cover change in Ban Mai Nong Bua and its vicinity, northern Thailand. Paper presented at the annual meeting of the American Anthropological Association, November 18, 2004, in San Francisco.

Reed, Richard K. 1990. Cultivating the tropical forest. In *Conformity and conflict: Readings in cultural anthropology,* eds. J. P. Spradley and D. W. McCurdy. New York: Harper Collins.

Reid, Anthony. 1996. Flows and seepages in the long-term Chinese interaction in Southeast Asia. In *Sojourners and settlers: Histories of Southeast Asian and the Chinese,* ed. Anthony Reid, 16–35. St. Leonards, Australia: Allen and Unwin.

Renard, Ronald. 1996. *The Burmese connection: Illegal drugs and the making of the Golden Triangle.* Boulder: Lynne Rienner Publishers.

Rerkasem, Kanok, Benjavan Rerkasem, Mingsarn Kaosa-ard, Chaiwat Roongruangsee, Sitanon Jesdapipat, Benchaphun Shinawatra, and Pornpen Wijukprasert. 1994. *Assessment of sustainable highland agricultural systems.* Bangkok, Thailand: Thailand Development Research Institute Foundation.

Reynolds, Craig J. 1996. Tycoons and warlords: Modern Thai social formations and Chinese historical romance. In *Sojourners and settlers: Histories of Southeast Asia and the Chinese,* ed. Anthony Reid, 115–147. St. Leonards, Australia: Allen & Unwin.

Roling, N. G., and M. A. E. Wagemakers. 2000. *Facilitating sustainable agriculture: Participatory learning and adaptative management in times of environmental uncertainty.* Cambridge: Cambridge University Press.

Salaff, Janet. 1975. *Working daughters of Hong Kong: Female filial piety or a transformation in the family power.* Toronto: University of Toronto Press

Santasombat, Yos. 2003. *Biodiversity, local knowledge, and sustainable development.* Chiang Mai: Regional Center for Social Science and Sustainable Development.

Sarna, Jonathan D. 1999. American Jewish political conservatism in historical perspective. *American Jewish History* 87: 113–122.

Sassen, Saskia. 1996. *Losing control? Sovereignty in an age of globalization.* New York: Columbia University Press.

Sassen, Saskia. 2007. Introduction: Deciphering the global. In *Deciphering the global: Its scales, spaces and subjects,* ed. Saskia Sassen. New York: Routledge.

Scott, James C. 1998. *Seeing like a state: How certain schemes to improve the human condition have failed.* New Haven: Yale University Press.

Sheffer, Garbriel. 1986. *Modern diaspora in international politics.* London: Croom Helm.

Sheffer, Garbriel. 1993. Ethnic diasporas: A threat to their hosts? In *International migration and security,* ed. Myron Weiner. Boulder: Westview Press.

Shigetomi, Shin'ichi. 1992. From "loosely" to "tightly" structured social organizations: The changing aspects of cooperation and village community in rural Thailand. *The Developing Economies,* 32: 154–178. Chiba, Japan: Developing Institute of Economies.

Skinner, G. William. 1957. *Chinese society in Thailand: An analytical history.* Ithaca, NY: Cornell University Press.

Skinner, G. William. 1996. Creolized Chinese societies in Southeast Asia. In *Sojourners and settlers: Histories of Southeast Asia and the Chinese,* ed. Anthony Reid, 51–94. St. Leonards, Australia: Allen & Unwin.

Solow, Robert. 1992. *An almost practical step toward sustainability.* Washington, DC: Resources for the Future Publications.

Stacey, Judith. 1983. *Patriarchy and socialist revolution in China.* Berkeley: University of California Press.

Stubbs, Paul. 1999. Virtual diasporas? Imagining Croatia on-line. *Sociological Research Online,* vol. 4, no. 2. http://www.socresonLine.or.uk/socresonLine/4/2/stubbs.html.

Sturgeon, Janet C. 2005. *Border landscapes: The politics of Akha land use in China and Thailand.* Seattle: University of Washington Press.

Szonyi, M. 1997. The illusion of standardizing the Gods: The cult of the five emperors in Late Imperial China. *The Journal of Asian Studies* 56:113–135.

Tam, Siumi Maria. 2006. Engendering Minnan mobility: Women sojourner in a patriarchal world. In *Southern Fujian: Reproduction of traditions in post-Mao China,* ed. Tan Chee-Beng. Hong Kong: The Chinese University Press.

Thongchai, Winichakul. 1992. *Siam mapped: A history of the geo-body of a nation.* Honolulu: University of Hawai'i Press.

Tu, Wei-ming. 1994. Ed. *Living tree: The changing meaning of being Chinese today.* Stanford: Stanford University Press.

Turner, Victor W. 1969. *The ritual process: Structure and anti-structure.* Chicago: Aldine Publishing Co.

Van Esterik, Penny. 2000. *Materializing Thailand.* Oxford and New York: Berg Publications.

Van Gennep, Arnold. 1960. *The rites of Passsge.* Translated by M. B. Vizedom and G. L. Caffee. Chicago: University of Chicago Press.

Van Hear, Nicholas. 1998. *New Diasporas: The mass exodus, dispersal and regrouping of migrant communities.* Seattle: University of Washington Press.

Walker, Andrew. 1999. *The legend of the Golden Boat: Regulation, trade and traders in the borderlands of Laos, Thailand, China and Burma.* Surrey, Great Britain: Curzon Press.

Wang, Gungwu. 1996. Sojourning: The Chinese experience in Southeast Asia. In *Sojourners and settlers: Histories of Southeast Asia and the Chinese,* ed. Anthony Reid, 1–15. St. Leonards, Australia: Allen & Unwin.

Watson, James. 1983. Standardizing the gods: The promotions of T'ien Hou ('Empress of Heaven') along the South China Coast, 960–1960. In *Popular culture in late Imperial China,* eds. D. Johnson, A. Nathan, and E. Rawski. Berkeley: University of California Press.

Watson, James. 1988. The structure of Chinese funerary rites: Elementary forms, ritual sequence, and the primacy of performance. In *Death ritual in later imperial and modern China,* eds. J. L. Watson and E. S. Rawski. Berkeley, Los Angeles: University of California Press.

Weiner, Myron. 1993. *International migration and security.* Boulder: Westview Press.

Weller, Robert. 1985. *Unities and diversities in Chinese religion.* London: Macmillan.

Wolf, Arthur P. 1969. Chinese kinship and mourning dress. In *Family and kinship in Chinese society,* ed. M. Freedman. Stanford: Stanford University Press.

Wolf, Arthur P. 1974. Gods, ghosts, and ancestors. In *Religion and ritual in Chinese society,* ed. A. P. Wolf. Stanford: Stanford University Press.

Wolf, Arthur P. 1985. Chinese family size: A myth revitalized. In *The Chinese family and Its ritual behavior,* eds. Hsieh and Chuang. Taipei: Institute of Ethnology, Academia Sinica Publications.

Wolf, Margery. 1972. *Women and the family in rural Taiwan.* Stanford: Stanford University Press.

Wolf, Margery. 1978. *The House of Lim: A study of a Chinese farm family.* Taipei: Caves Books.

Wolf, Margery. 1985. *Revolution postponed: Women in contemporary China.* Stanford: Stanford University Press.

Wright, P. C. 1993. Ranomafana National Park, Madagascar: Rainforest conservation and economic development. Paper prepared for Liz Claiborne—Art Ortenberg Foundation Community-Based Conservation Workshop, Arlie, Virginia.

Wyatt, David K. 1982. *Thailand: A short history.* New Haven: Yale University Press.

Yan, Yunxiang. 1996. *The flow of gifts: Reciprocity and social network in a Chinese village.* Stanford: Stanford University Press.

Yeoh, Brenda S. A., Michael W. Charney, and Tong Chee Kiong. 2003. Eds. *Approaching transnationalisms: Studies on transnational societies, multicultural contacts, and imaginings of home.* Boston: Kluwer Academia Publishers.

Young, Anthony. 1998. *Land resources: Now and for the future.* Cambridge: University Press.

Young, Gordon. 1962. *The hill tribes of Northern Thailand (A Socio-Ethnological Report).* New York: AMS Press.

Index